Monographs on Astronomical Subjects

General Editor, A. J. Meadows, D.Phil.,
Professor of Astronomy, University of Leicester

The Origin of the Planets

To be published in the same series:

Magnetohydrodynamics
T. G. Cowling, M.A., D.Phil., F.R.S.

Theories of Gravitation and Geophysics
P. S. Wesson, Ph.D.

Interstellar Clouds
D. McNally, Ph.D.

X-ray and Gamma-Ray Astronomy
David Adams, Ph.D.

The Interpretation of Solar Flares
P. A. Sweet, Ph.D.

Solar X-ray Astronomy
J. H. Parkinson, Ph.D.

Solar Radio Emission
J. F. Vesecky, Ph.D.

Planetary Atmospheres
G. E. Hunt, Ph.D.

Orbital Motion
A. E. Roy, Ph.D.

Resonances in the Solar System
S. F. Dermott, Ph.D.

The Origin of the Planets

I. P. Williams, Ph.D.

Reader in Applied Mathematics, Queen Mary College
University of London

First Published September 1975

© I. P. Williams 1975

All rights reserved. No part of this publication may be reproduced, stored in a retrieval system, or transmitted, in any form or by any means, electronic, mechanical, photocopying, recording or otherwise, without the prior permission of Adam Hilger Limited.

ISBN 0 85274 258 4

Published by Adam Hilger

A company now owned by
The Institute of Physics
Techno House
Redcliffe Way
Bristol BS1 6NX
England

Contents

Preface vii

1. **The Planetary System** 1
   1.1 Introduction 1
   1.2 Other Planetary Systems 2
   1.3 The Sun 6
   1.4 The Planets 14
   1.5 The Satellite System 18
   1.6 The Asteroids 20
   1.7 Comets 22
   1.8 Meteorites 23
   1.9 The Main Features of the Planetary System ... 25
   1.10 Classification of Theories 27

2. **The Tidal and Other Related Theories** 29
   2.1 Introduction 29
   2.2 The Tidal Theories of Jeans and Jeffreys 31
   2.3 Objections to the Tidal Theories 38
   2.4 Theories Involving Multiple Stars 42
   2.5 An Unusual Theory 45
   2.6 Woolfson's Theory of Planetary Formation ... 46

3. **The Accretion Theories** 51
   3.1 Introduction 51
   3.2 Schmidt's Theory 54
   3.3 Alfvén's Theory 56
   3.4 The Theory of Pendred and Williams 60

4. **Processes for the Formation of the Sun and Planets** ... 65
   4.1 Introduction 65
   4.2 Descartes' Theory and its Successors 66
   4.3 The Theory of Kant and its Successors 79
   4.4 The Theory of Laplace and its Successors ... 85

5.	**Summary and Conclusions**	89
	5.1 Introduction	89
	5.2 The Main Theories for the Formation of the Planets	91
	5.3 Other Relevant Information	95
	5.4 Concluding Remarks	96

References 99

Index 105

Preface

The origin of the planets is a topic of great fascination to mankind and in consequence many a theory has been proposed to explain it. This book attempts to review such theories in slightly more depth than a qualitative review paper. To this end, some of the mathematical formulation of various theories has been included, though it is hoped that a general reader might be able to omit such calculations and still obtain some insight into the theories.

I should like to thank all my colleagues at Queen Mary College for many stimulating discussions and to thank Miss J. Percival and Mrs. L. Buer for assistance with the manuscript.

I. P. Williams

1. *The Planetary System*

1.1. Introduction

The formation and evolution of the planetary system have been a source of fascination to both laymen and professional astronomers for many centuries. Of course, pre-Copernican theories were hampered by the belief that the Earth was the centre of the whole Universe and therefore that the formation of the Earth was naturally related to that of the Universe. We shall not concern ourselves with a discussion of such theories here, even though some of them may be of considerable historic interest. Instead, we shall deal only with theories where the centre of the system is recognized to be the Sun and where formation occurs from matter which already exists in a state which is familiar to us (that is, as other stars, interstellar gas clouds or some similar objects).

There are two great difficulties facing any prospective theorist in this subject: one which is common to all formation problems and one which is peculiar to this problem. The first difficulty is that we are discussing the evolution of the system from an unknown initial state to a known final (present) state. Since most of the processes envisaged are not time-reversible, then this is not a simple question of reversing time and working backwards. Instead we have to postulate a reasonable initial state and demonstrate that the final state can be obtained from this. Unfortunately, there may be many paths which lead to a situation resembling our given final state; this partly accounts for the multiplicity of theories that exist for the origin of the planets. This problem of the non-uniqueness of the path to the final state can be illustrated as follows. If a man with a parachute jumps from an aeroplane at some given height, then it is simply a matter of aerodynamics to calculate where and at what height he will be at any subsequent time. If, on the other hand, all we can do is to observe this parachutist falling past our window, it is impossible to determine whether he came from the roof of the building or from an aeroplane, not to mention calculating exactly when and at what height he jumped.

The second problem arises because we have detailed information about only one planetary system, namely our own. This is in fact a twin problem. If we are indeed the only planetary system in existence, then we can say very little about our origin for the normal rules of statistics cannot be used on unique events and any event, however improbable, can come about once. This may be illustrated by the fate of the elephant at Berlin Zoo. During the last war, Berlin Zoo had one elephant. The probability of this elephant being killed in the first bombing raid on Berlin must be infinitesimally small yet killed it was. Similarly any event, however unlikely, could have occurred once and may therefore have played a part in the formation of a unique planetary system.

The second branch of this problem arises if our planetary system is not unique. Then we have to deduce, from observations of our own system, what the general characteristics of planetary systems are and which properties are essentially peculiar to our system. As an illustration of this problem consider the situation where some hypothetical Xian from planet X has succeeded in capturing and taking home with him one human being, say the editor of this series. From observations of this one specimen, the Xian is attempting to categorize the main features of the human race. With careful thought the Xian might deduce that humans have two legs, two arms and so on. He would, however, be on dangerous grounds if he tried to deduce anything about the colouring of the eyes, never mind the colour and quantity of hair possessed by the average human being. This illustration should be borne in mind whenever we attempt to categorize the properties of planetary systems.

1.2. Other Planetary Systems

Of course, it would be of considerable help if another planetary system could be discovered and so it seems appropriate to give an account of the work that is being carried out in attempting to locate such a system.

Unfortunately, existing telescopes are not powerful enough to directly detect planets orbiting even the nearest stars. This is because planets essentially shine by reflected light from their parent star. If the planet is far away from its parent, it becomes too faint to be seen. But if it is close to its parent, so that it becomes brighter, it is swamped by the light from the star itself. As an example consider our own system. From α Centauri, the Sun would appear as a star of apparent magnitude +0.3. Jupiter would be a 23rd magnitude star. Such a faint object would hardly be detectable by the 200 inch Mount Palomar telescope even without a bright star a few seconds of arc away. We are therefore forced to use indirect means if we hope to detect other planetary systems.

The most obvious method involves the careful study of the proper motion of nearby stars. Since it is the centre of mass of any system that moves on a well defined curve, an individual component in a system consisting of more than one component would show an irregular proper motion. Conversely, if a star is found that has a very irregular proper motion, it is very probably part of a multiple system and with very careful observation of these irregularities it may be possible to deduce the mass and mean orbital radius of the components. Of course, most stars showing this behaviour will be components of binary systems and will be of no interest in the present context. If it is a planet that is to be discovered, then it effectively contributes nothing towards the total light output and its mass is at the most comparable to that of Jupiter. Call the mass of the star and an orbiting planet M_S and M_P respectively; then if a_S and a_P are the semi-major axes of the orbits of the star and planet about the centre of mass of the system, Kepler's third law yields

$$P^2 = \frac{4\pi^2 a^3}{G(M_P + M_S)}, \qquad (1.1)$$

where P is the period and $a = a_P + a_S$. However, by the definition of the centre of mass,

$$a = \frac{a_S}{M_P}(M_P + M_S), \qquad (1.2)$$

so that from equation (1.1),

$$\frac{P^2}{a_S^3} = \left(\frac{M_P + M_S}{M_P}\right)^3 \cdot \frac{4\pi^2}{G(M_P + M_S)}. \qquad (1.3)$$

From observations of the proper motion, a_S and P can be determined, which leaves one equation in two unknowns. If, however, we know the mass of the star, or can make a reasonable estimate of it from its luminosity, then we can obtain M_P, the mass of the planet. (It should be remembered that the parallax of every star for which the above method might work will be known, so that the luminosity will be easily obtained from the apparent magnitude).

The most successful and consistent user of the above method has been van de Kamp at Sproul Observatory. From van de Kamp's work (1969a, b; 1971), Table 1.1 can be constructed showing those nearby stars that possess unseen companions, possibly of a planetary nature, together with information regarding the parent star.

We see that there are five stars, Barnard's star, Lalande 21185, ε Eridani, 61 Cygni and AO_e 17415–6 that might have planetary companions. Barnard's star deserves further mention in that Black & Suffolk (1973) have also analysed van de Kamp's observations, and have concluded that it is a multiple planet system in which the planets may not be coplanar. Perhaps, however, this result should be regarded with some reserve at present.

There are two other indirect methods which will, in theory, allow the discovery of a planet or planetary system in orbit about a star to be made. If, by chance, we on Earth lie close to the orbital plane of another planetary system, then periodically we should be able to observe the transit of a planet across the face of its parent star. We cannot, of course, resolve an individual planet on the face of a nearby star, but we should observe a

Table 1.1 The Data for Nearby Stars Suspected of Having Planets (after van de Kamp, 1969a, b; 1971)

Name of Star	R. A. (1950) h m	Dec. (1950) °	Distance Away (light years)	Mass of Parent Star (solar masses)	Suspected Mass of Planet (Jupiter masses)	Comments
Barnard's Star	17 55.2	+ 4 33	5.9	0.15	1.1 0.8	Also analysis by Black & Suffolk (1973) suggesting 3 planets with masses similar to those given.
Lalande 21185	11 00.6	+36 18	8.1		10	Regard with considerable reserve.
ε Eridani	3 30.6	− 9 38	10.7		6	
61 Cygni	21 04.7	+38 30	11.2	0.58 0.57	8	Unseen companion is attached to the first of the two stars in this binary.
AO$_e$174156	17 36.7	+68 23	15.7		25	Believed to have a highly eccentric orbit. Rather high mass for a planet.

change in the luminosity of the star while transit is occurring. For example, the transit of Jupiter viewed from outside our system would produce a drop in the apparent luminosity of the Sun by about 1 per cent. Such an event would occur every 12 years, last about one day, and the observer would have to be within 0·2° of the plane of the orbit of Jupiter to observe it. There is, therefore, a small but finite probability of discovering a planet by such a method, though no planets have been discovered using this technique so far.

The viability of the other method depends on the fact that residual gases and dust grains within the Solar system scatter more light from the Sun than any planet. It is, therefore, easier to discover such a halo of scattered light than it is to pick up the light reflected by single planets. We could search the sky for stars showing such haloes, and if one were to be discovered, we could then determine the orbital plane of the planetary system from its elongation (though nothing could be discovered about the individual planets). Unfortunately, no planetary systems have yet been discovered by this technique either.

What evidence there is does suggest that our planetary system is not unique, but there is not enough information available to be of any use in compiling a list of the major properties of general planetary systems, so we are forced to use information about our systems alone. We will first describe each of the main components of our system, and will then compile a list of its major properties.

1.3. The Sun

Most of the material in the solar system resides in the Sun, its mass being $1\cdot991 \times 10^{30}$ kg. The Sun rotates fairly slowly, the angular velocity at its equator being only 2.90×10^{-6} rad s^{-1}. This, combined with a radius of $6\cdot960 \times 10^8$ m, leads to a moment of inertia of 5×10^{46} kg m^2. Its angular momentum, if it rotates as a rigid body, is therefore $1\cdot5 \times 10^{41}$ kg m^2 s^{-1}, which is considerably less than the sum of the angular momenta of the planets, namely $3\cdot15 \times 10^{43}$ kg m^2 s^{-1}.

At present, the Sun is a normal main sequence star of spectral type GIV. It has a composition roughly similar to interstellar clouds, namely 60–70 per cent hydrogen, 30–40 per cent helium and under 3 per cent other elements, and generates energy by the conversion of hydrogen to helium. Its luminosity is 3.86×10^{26} J s^{-1} and this fact, together with the given value for the radius, means that its effective temperature is 5785 K. A black body placed at a distance r (measured in units of 10^{11} m) from the Sun would achieve an equilibrium temperature of about $340/(r)^{1/2}$ K.

It is known that corpuscular radiation, called the 'solar wind', flows outward from the Sun. Its strength varies, but it has typical densities of about 5×10^6 atom m^{-3} and a velocity of 4.5×10^5 m s^{-1} at the radius of the Earth's orbit. It therefore represents a mass loss of over 10^8 kg s^{-1} from the Sun, but this amount is insignificant compared with the mass of the Sun. One interesting feature of the wind is that it appears to be able to carry the solar magnetic field lines with it, so that these lines reach far out from the Sun in a long spiral. It is widely believed that this allows a considerable transfer of angular momentum to take place from the Sun to interstellar space, and so accounts for the low rotational velocity of the Sun. The magnetic field has a strength of about 200 ampere turns m^{-1} at the Sun's pole. All the above data relating to the Sun have been taken from Allen (1973).

So far as the problem of the formation and evolution of planets is concerned, the possible past behaviour of the Sun may be of more importance than its present character. We shall, therefore, investigate this in some detail. Unfortunately, it is not yet known in detail how stars form, but later we shall discuss briefly some of the theories, at least in so far as they relate to the formation of planets. It is generally accepted, however, that before the Sun (or any other star) reached the hydrogen-burning stage on the main sequence, its energy was provided by the release of gravitational energy because of its contraction. For any star where the transport of energy in its interior is by radiation, the radiative flux is given by

$$F_r = -\frac{4acT^3}{3\kappa\rho}\frac{dT}{dr} \qquad (1.4)$$

where T is the temperature, ρ the density, $ac/4$ the Stefan–Boltzmann constant and κ the opacity. (For more detail on this, and on what follows, see e.g. Schwarzschild, 1958). Very approximately Kramer's opacity law applies, so that

$$\kappa = \kappa_0 \rho T^{-3.5} \tag{1.5}$$

where κ_0 is a constant. If the star can be treated as a polytrope (that is, $P = C\rho^n$), then $\rho \sim M/R^3$ and $T \sim M/R$, which gives

$$F_r \sim M^{5.5}/R^{2.5}. \tag{1.6}$$

But the luminosity of the star is

$$L = 4\pi R^2 F_r, \tag{1.7}$$

and

$$L = 4\pi R^2 \frac{ac}{4} T_{\text{eff}}^4, \tag{1.8}$$

Hence $L \sim M^{5.5}/R^{0.5}$ and $T_{\text{eff}} \sim M^{5.5}/R^{2.5}$ so that the track in the HR diagram ($\log L$ plotted against $\log T_{\text{eff}}$) of a given mass is the straight line

$$\log L = 0.2 \log T_{\text{eff}} + \text{Constant}. \tag{1.9}$$

The main feature of this type of evolution is that at large radius, both the luminosity and the effective temperature are lower than at present: the calculated track for a star of solar mass is shown in Figure 1.1. Of course, more detailed stellar models present some variation from this simple picture (e.g. see Henyey et al., 1955), but it displays the main features of interest, namely a trajectory of increasing temperature and of much more slowly increasing luminosity.

Unfortunately, if the surface temperature was at one stage low, as is implied by the above argument, the opacity would no longer have been given by Kramer's law but is instead generated by negative hydrogen ions. The general expression for κ is then rather complex (see Cox & Guili, 1968) but an approximate simple power law fit is given by

$$\kappa = \kappa_1 P^{1/2} T^4 \tag{1.10}$$

where κ_1 is a known constant.

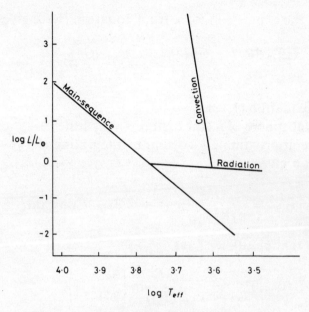

Figure 1.1 Schematic HR diagram for young stars.

For any work on the transfer of radiation, it is convenient to define the optical depth, τ, by

$$\tau = \int_r^\infty \kappa\rho \, dr. \qquad (1.11)$$

With this definition, it is easy to obtain the temperature in the atmosphere of a star as

$$T^4 = \tfrac{3}{4} T_{\text{eff}}^4 (\tau + \tfrac{2}{3}). \qquad (1.12)$$

The atmosphere must be supported against gravity by the pressure gradient, and so

$$\frac{1}{\rho} \frac{dP}{dr} = -\frac{GM}{R^2},$$

or, making use of the optical depth τ,

$$\frac{dP}{d\tau} = \frac{GM}{\kappa R^2}. \qquad (1.13)$$

Inserting the expression for κ from equation (1.10) gives

$$P^{1/2}\frac{dP}{d\tau} = \frac{GM}{R^2\kappa_1 T^4} = \frac{4GM}{3R^2\kappa_1 T_{\text{eff}}^4(\tau + 2/3)} \qquad (1.14)$$

on using equation (1.12).

The atmosphere of a star contains very little mass, and so M and R are approximately constant. Integrating (1.14) with this assumption gives

$$P^{3/2} = \frac{2GM}{\kappa_1 R^2 T_{\text{eff}}^4}\ln(\tau + 2/3) + \text{Constant}, \qquad (1.15)$$

or, if $P = 0$ at $\tau = 0$, we have

$$P^{3/2} = \frac{2GM}{\kappa_1 R^2 T_{\text{eff}}^4}\ln\left(\frac{3}{2}\tau + 1\right). \qquad (1.16)$$

Now, a star becomes convective if

$$\frac{T\,dP}{P\,dT} < \frac{5}{2}$$

and by substituting from equations (1.12), (1.13) and (1.16), we find that convection will occur provided

$$\ln\left(\tfrac{3}{2}\tau + 1\right) > 16/15. \qquad (1.17)$$

This gives the perfectly reasonable value of 1·267 for τ and so we must conclude that when the opacity law for the Sun is due to negative hydrogen ions it will be convective, thus creating the drastic change to its constitution mentioned earlier. The main feature of any convective star is that

$$P \sim T^{5/2},$$

that is, it behaves like a polytrope of index 3/2. We therefore have

$$\frac{P}{P_c} = \left(\frac{T}{T_c}\right)^{5/2}, \qquad (1.18)$$

where P_c and T_c are the central pressure and temperature, respectively, and are given by

$$P_c = A\frac{GM^2}{R^4},$$

$$T_c = B\frac{M}{R},$$

where A and B are known numerical constants. Therefore throughout the convective layer

$$P = CM^{-1/2}R^{-3/2}T^{5/2}, \qquad (1.19)$$

where C is a constant related to A and B. At the boundary between atmosphere and star, this relationship between P, R and T must match that given for the atmosphere by equation (1.16) with $\tau = 1\cdot267$. This requires that

$$T_{\text{eff}} = C_1\left(\frac{M}{M_\odot}\right)^{7/31}\left(\frac{R}{R_\odot}\right)^{1/31} \qquad (1.20)$$

where C_1 is a constant whose numerical value is about 3500 K.

Using the black-body relationship, we have

$$L = D\left(\frac{M}{M_\odot}\right)^{28/31}\left(\frac{R}{R_\odot}\right)^{66/31}, \qquad (1.21)$$

where D is another constant with a known value. Combining equations (1.20) and (1.21) gives the relationship between luminosity and temperature as

$$\log L = 66 \log T_{\text{eff}} + \text{Constant.} \qquad (1.22)$$

This equation reveals the drastic changes in the star's behaviour that must be expected: since L decreases as R decreases, the luminosity must have been much greater in the past. The calculated track in the HR diagram becomes a line along which the luminosity changes rapidly, but there is hardly any change in effective temperature. This track is also shown in Figure 1.1. Detailed calculations by Hayashi (1961), Faulkner *et al.* (1963), Ezer & Cameron (1963) and others, have shown that the main features of this picture are correct.

The point of interest so far as planetary cosmogony is concerned is that the luminosity and radius were higher in the past. This means that the equilibrium temperature of any body near the Sun would have been increased by a factor $(L/L_\odot)^{1/4}$ from its present day value of $340/(r)^{1/2}$ K. This can obviously be of considerable importance in any situation where the condensation of materials is under discussion. A second point to note is that, as the radius was higher in the past, the Sun occupied a larger volume, and this fact alone could account for the absence of planets within the orbit of Mercury.

Another feature that may have been enhanced in this early phase was the solar wind. We may see this very simply as follows (Williams, 1967). If the rate of mass loss is m, then energy is carried away by this mass at least at the rate GMm/R, since the material must move with at least the escape velocity from the star. But the only source of energy readily available is the Sun's own luminosity, L and so we might expect

$$m \sim \frac{RL}{GM}. \qquad (1.23)$$

This indicates a much enhanced wind when R and L are both large. Observation of T-Tauri stars (Kuhi, 1964) gives support to this belief, and indicates that a mass-loss rate of the order of $10^{-6} M_\odot$ yr^{-1} (10^{17} kg s^{-1}) is possible. The solar wind may therefore have a significant effect in the early stages of planetary formation, since it might then be able to carry away any residual gas.

It is therefore important to estimate the length of time for which the high-luminosity, high-radius stage might last. In this early stage, the Sun is completely dependent on its own gravitational potential energy as a source of energy. For a convective star with a polytropic index of 3/2, this is given by

$$-\frac{6}{7}\frac{GM^2}{R}, \qquad (1.24)$$

(see Cox & Giuli, 1968). If we assume that the Sun is composed of an adiabatic gas, with $\gamma = 5/3$, then it can be shown that half of the energy released in contraction can be radiated away while the other half is used to heat up the solar interior. The relationship between energy available and energy radiated is therefore

$$L = -\frac{d}{dt}\left(\frac{3}{7}\frac{GM^2}{R}\right).$$

But equation (1.21) gives

$$L = D\left(\frac{M}{M_\odot}\right)^{28/31}\left(\frac{R}{R_\odot}\right)^{66/31}$$

and so the above equation becomes

$$DR^{128/31} = -\tfrac{3}{7}GM_\odot^{28/31}M^{34/31}R_\odot^{66/31}\frac{dR}{dt}.$$

Since M is roughly constant, the above equation can be integrated to give

$$\begin{aligned}t &= \frac{3}{7D}GM_\odot^{28/31}M^{34/31}R_\odot^{66/31} \cdot \frac{1}{R^{97/31}}\frac{31}{97} \\ &= \frac{93GM^2}{679RL},\end{aligned} \quad (1.25)$$

assuming that $t = 0$ when R is large.

If the present-day values of M, R and L are inserted into equation (1.25), then the contraction time is about 4×10^6 years. The high-luminosity, high-solar-wind phase does not, therefore, last very long, and so planets must be in evidence very soon after the Sun is formed if this phase is to be an important factor in their development.

This concludes our brief account of the Sun as a star. Of course none, or only selected parts, of the foregoing may prove to be relevant. Different theories make use of different aspects of the Sun's behaviour.

1.4. The Planets

It is the existence and observability of planets that distinguishes our system from any other known system, and the problem of their origin must form a main part of this treatise. There are nine known planets in the system at present (the host of minor objects will be mentioned later), and their main dynamical characteristics are given in Table 1.2. There exists an empirical law, the Titius–Bode law (see Nieto, 1973) which gives the mean orbital distances of the planets from the Sun. One form of the law can be written

$$r = 0.6 + 0.45 \times 2^n \qquad (1.26)$$

where $n = -\infty$ for Mercury, 0 for Venus, 1 for the Earth and so on. The asteroidal belt must be included in this sequence, and r is measured in units of 10^{11} m. The orbital distance as calculated from the Titius–Bode law is also included in Table 1.2, for comparison with the actual orbital distances.

From Table 1.2, it is immediately evident that all the planets move in nearly circular orbits, and that these orbits are very nearly coplanar. What is not shown in the table, but is nevertheless true, is that all the planets move in the same, prograde, sense about their orbit, so that the vectors associated with the angular momentum of the planets are nearly parallel to one another. Jupiter has the angular momentum largest in magnitude relative to the Sun, namely 2×10^{43} kg m s^{-1}, followed by Saturn, Neptune and Uranus, in that order.

There is also a tendency for planets to rotate in a prograde fashion on an axis perpendicular to their orbital plane through their own mass centre, but this is by no means so universal: Venus, Uranus and Pluto are clear exceptions. The relevant information is given in Table 1.3.

There is considerable current interest in the interiors of planets; that is, in calculating equilibrium states and deducing information regarding the state of the interior of planets. Such research has received a major impetus from the recent growth of spaceflight. This is such a vast field that we will make no attempt to cover it here. The particular part of this work that seems of

Table 1.2 Dynamical Characteristics of the Planetary Orbits (Data from Allen, 1973)

Planet	Mass (10^{24} kg)	Inclination to ecliptic (i)	Eccentricity (e)	Mean Orbital distance (10^{11} m)	Titius–Bode's Law distance (10^{11} m)
Mercury	0.33	7° 0'	0.2056	0.58	0.60
Venus	4.9	3°24'	0.0068	1.08	1.05
Earth	6.0	—	0.0167	1.50	1.50
Mars	0.64	1°51'	0.0934	2.28	2.40
Asteroids	—	—	—	4.35	4.20
Jupiter	1900	1°18'	0.0485	7.78	7.80
Saturn	570	2°29'	0.0557	14.24	15.00
Uranus	87	0°46'	0.0472	28.70	29.40
Neptune	100	1°46'	0.0086	44.97	58.20
Pluto	1.0	17°10'	0.250	59.00	115.80

Table 1.3 Rotation of the Planets (Data from Allen, 1973)

Planet	Inclination of Equator to orbit	Rotation Period d	h	m	Moment of Inertia (Earth Units)
Mercury	<28°	59			1·20
Venus	187°	244	8		1·02
Earth	23°27'		23	56	1·00
Mars	23°59'		24	37	1·13
Jupiter	3°05'	1	9	50	0·75
Saturn	26°44'		10	14	0·66
Uranus	97°55'		10	49	0·69
Neptune	28°48'		15	48	0·87
Pluto	(90°)?	(6	9)?		?

direct value in attempting to understand the formation of the planets is that which provides information on the composition of the planets.

For most of the planets, information regarding the surface layers can be obtained either from an analysis of the reflected radiation, or from a direct study of samples. These outer layers, however, contain only a fraction of the mass, and if we wish to talk about the composition of a planet, then we must have data concerning its interior. One piece of information which helps us to deduce what the interior composition might be is the mean density. We can also deduce the moment of inertia of most of the planets, and this gives a strong indication of how centrally condensed they are. Because materials in the interior of planets are much more heavily compressed, their densities are much higher than they would be on Earth and it is easy to demonstrate that Jupiter with a density of $1·33 \times 10^3$ kg m^{-3} cannot contain an appreciable quantity of any material heavier than hydrogen or helium. The same appears to be true for Saturn. For other planets, however, the mean density is somewhat higher and we cannot uniquely determine the composition from these considerations alone, since there is no way to distinguish between a planet composed of, say, half hydrogen and half iron and one composed entirely of whatever element has a mass equal to the

Table 1.4 Mass and Composition of the Planets

Type	Planet	Orbital Distance (10^{11} m)	Mass (10^{24} kg)	Composition
Terrestrial	Venus	1·08	4·9	Fe, Si, O
	Earth	1·50	6·0	(Refractory Type)
Major	Jupiter	7·78	1900	H, He
	Saturn	14·24	570	
Outer	Uranus	28·70	87	C, N, O
	Neptune	44·97	100	H, He
Others	Mercury	0·58	0·33	Fe, Si, O?
	Mars	2·28	0·64	
	Pluto	59·00	1·00	
Satellites	Ganymede	7·78	0·15	Fe, Si, O,
	Triton	44·97	0·15	various ices
	Titan	14·27	0·14	
	Callisto	7·78	0·09	
	Io	7·78	0·08	
	Moon	1·50	0·07	

average of hydrogen and iron. If we also assume that the cosmically more abundant elements are also likely to be the most abundant in the planets, then a reasonable estimate for the composition of the planets can be made. It transpires that the planets can be divided into three subgroups, each subgroup differing in mass, composition and position relative to the Sun. This information is given in Table 1.4. It is also possible that a fourth subgroup exists—namely, escaped satellites. For comparison, some of the larger satellites have therefore been included in Table 1.4. Information regarding the magnetic fields of planets tends to confirm the picture we have, in that it indicates the presence of a metallic core in those planets we would expect on other evidence to possess such a core.

There is also a division within the terrestrial planets themselves, basically between those planets which are iron-rich (the Earth) and those deficient in iron (Mars and the Moon).

Mercury may also be iron-rich, but Venus is fairly normal. This fractionation of the iron and the silicates could occur while the refractory material is condensing, during an accumulation state or afterwards. Meteorites are found to be similarly segregated, and so this may be a widespread process.

There are, of course, many other properties of individual planets that could be mentioned here, but what has been given essentially summarizes the information directly relevant to a discussion of the origin and evolution of the solar system.

1.5. The Satellite Systems

We are naturally familiar with the existence of the single satellite of the Earth, namely the Moon. More extensive satellite systems surround some of the planets in the solar system. Table 1.5 lists the main dynamical properties of these satellites. We can see from the table that in some senses the systems around Jupiter, Saturn and Uranus could be regarded as mini-planetary systems. But there are also important differences. Firstly, the mass ratio of satellite to planet is much smaller than the corresponding ratio of planet to Sun for the vast majority of the satellites. Secondly, a fair proportion of the satellites have retrograde orbits and their general alignment to a plane is not so good as the mean alignment of the planetary system. Finally, of course, the whole satellite system is on a much smaller scale. It should be noted, however, that the satellite systems satisfy a Titius–Bode type law rather more accurately than the planetary system itself (see Dermott, 1972). There is a host of interesting questions concerned with individual satellites or satellite systems. One typical question of this nature is the possible existence of atmospheres on some of the larger satellites, in particular Io, Europa, Ganymede and Callisto. Another question of interest is the relationship, if any, between the satellite system and ring system surrounding Saturn. Such problems, while of considerable general interest, are not necessarily of direct relevance to the central question which we are considering, namely the formation of the solar system. In this context the interest in the satellites

Table 1.5 The Satellite Systems (Data from Allen, 1973)

Parent		Satellite	Distance from Parent (10^8 m)	Mass (10^{21} kg)	Orbit Inclination	Orbit Eccentricity
Earth		Moon	384	73·5	23	0·055
Mars	1	Phobos	9		1	0·021
	2	Deimos	23		2	0·003
Jupiter	1	Io	422	73	0	0·000
	2	Europa	671	48	1	0·000
	3	Ganymede	1070	154	0	0·001
	4	Callisto	1883	95	0	0·007
	5		181		0	0·003
	6		11476		28	0·158
	7		11737		26	0·207
	8		23500		147	0·40
	9		23600		156	0·275
	10		11700		29	0·12
	11		22600		163	0·207
	12		21200		147	0·169
Saturn	1	Mimas	186	0·04	2	0·020
	2	Enceladus	238	0·03	0	0·004
	3	Tethys	295	0·64	1	0·000
	4	Dione	377	1·1	0	0·002
	5	Rhea	527	2·3	0	0·001
	6	Titan	1222	137·0	0	0·029
	7	Hyperon	1483	0·1	1	0·104
	8	Iapetus	3560	1·1	15	0·028
	9	Phoebe	12950		150	0·163
	10	Janus	159		0	0·0
Uranus	1	Ariel	192	1·3	0	0·003
	2	Umbriel	267	0·5	0	0·004
	3	Titania	438	4·3	0	0·002
	4	Oberon	586	2·6	0	0·001
	5	Miranda	130	0·1	0	0·00
Neptune	1	Triton	355	140·0	160	0·00
	2	Nereid	5562		28	0·75

is centred on the question: did the satellite systems form in the same manner as the planetary system, only on a much smaller scale? If we answer this question in the negative, then for a complete theory we must propose some other means by which the satellites formed.

One must confess that, in general, theories for the formation of the planets tend to ignore the satellites completely, either saying that the process of planetary formation is repeated for the satellites on a smaller scale, or simply dismissing them by saying that the satellites formed after the planets had formed, and therefore that their formation is of no consequence for the formation of the planets. This latter argument is perhaps justified, but one is left with the feeling that, aesthetically at least, a theory actually capable of explaining the satellites as well as the planets is much to be preferred.

One last point which should be made is that the larger satellites are the third most massive class of objects in the solar system after the Sun and the planets; but the mass distribution of the minor planets, which we discuss next, follows fairly continuously after that of the smaller satellites.

1.6. The Asteroids

It is by now well known that there exists a whole family of objects whose orbits lie mainly between those of Mars and Jupiter. The larger of these minor planets have been given names, and Ceres, the largest, is often used in Titius–Bode's law to fill the orbital distance for $n = 3$. The total mass of the whole system of asteroids is unlikely to be more than about 10^{22} kg, while the largest individual mass is 6×10^{20} kg for Ceres.

The mean inclination of the whole system is about 9° to the ecliptic, though certain asteroids have a much higher inclination, for example, Pallas is inclined at some 35°. There is also a considerable variation in the eccentricity of asteroidal orbits.

Asteroids are affected by the gravitational field of Jupiter, and in consequence, conspicuous gaps are to be found corresponding to orbits with periods 1/3, 2/5 and 1/2 of Jupiter's orbital period.

The main interest in the system from the point of view of planetary cosmogony concerns the origin of the asteroidal belt: whether it represents an earlier stage in the evolution of a planet, or the state after some cataclysmic event in a planet's history, or whether it results from some other process altogether. The general feeling is that the last possibility is most likely to be correct: that is that the belt is neither a planet in the making, nor a planet after destruction. The main reason for this conclusion is the obvious one that the total mass of the asteroidal belt is much less than that of a single planet, and that the energy required to drive away the difference between a planetary mass and the asteroidal mass is much greater than that which can be supplied by any obvious source. However, it is possible that some of the same mechanisms now at play in the asteroidal belt were active in the formation stage of the planets: for example, small asteroids could be adhering together to form larger bodies. If this were found to be the case, it would lend strong support to any theory of planetary formation which makes use of a similar process. Unfortunately, though we cannot pretend to understand fully how the asteroidal belt evolves, all the indications suggest that the asteroidal belt has formed from a few larger (but much less than planetary-sized) objects, as a consequence of collisions leading to fragmentation (see e.g. Hellyer, 1970). This does not, of course, rule out the possibility of planets forming by the aggregation of small bodies. Indeed, it is possible that the initial large asteroids were formed by such aggregations (when there was considerable gas around, for example), and that it is during a later stage of development of the system that fragmentation occurs (when the gas has evaporated perhaps). Our present level of knowledge concerning the asteroidal belt does not therefore provide any direct lead to the formation of planets.

Since there are a number of ways in which the belt could have formed, existence does not place any restrictions (over and above any placed by the planets themselves) on the type of theory that is acceptable for the formation of the solar system.

1.7. Comets

Comets have always been associated with the mysterious and the supernatural in the minds of men, and even today our knowledge of them is very limited. Their main features are easy to describe. Most move on nearly parabolic orbits, so that they spend most of their life far away from the Sun, only coming close to it for a short interval during perihelion passage. The inclination of their orbits to the plane of the ecliptic and the differences between these inclinations are much greater than for any of the other solar system members that we have discussed. When the comet is close to perihelion, it has a large head which can have dimensions in excess of those of a planet. But the mass is much smaller—of the order of only 10^{15} kg—which gives the head a mean density of the order of 10^{-9} kg m^{-3}. The main feature of a comet is its tail, which always sweeps away from the Sun, the obvious cause being a combination of radiation pressure from the Sun and the solar wind.

There is considerable uncertainty about both the structure and the origin of comets (see Lyttleton, 1956; Richter, 1963). Since new comets are continually being discovered and it is known that some old comets are lost, one school of thought maintains that comets must form after the planetary system and that this formation process still operates from time to time, even to this day. If this view is correct, then comets have nothing to do with the origin of the solar system, and so are of no concern to us in this discussion.

The alternative view is that comets were formed at the same time as the planets, though of course much further away from the Sun, and that most comets still reside in the outer regions of the solar system. A few are perturbed from time to time by the action of the stars, to be discovered by us as new comets. On this view, comets are essentially composed of the materials left over from the formation of the planets. Since most theories for the origin of planets are in some way inefficient, all theories predict a large amount of debris, and so the existence of comets, even on the view that they are related to the planets in their formation, hardly places any restrictions on theory. It would be useful, if this view is correct, to obtain samples from a comet for inspection, since this would then allow an investigation of some of the debris left over from planetary formation and information about its composition would be of considerable interest. We have to wait, either for a rocket trip to a comet or for a comet to come to us, before such an investigation can be carried out, though of course spectroscopic analysis can give some information. Such investigations show that by far the most abundant elements are hydrogen, carbon, oxygen and nitrogen, occurring in a variety of molecular forms.

1.8. Meteorites

It is generally accepted that the meteorites originate from the asteroidal belt, or from some other primitive body that was present in the inner solar system. Thus they may represent matter which is in a state fairly similar to that which existed when the planets formed, at least so far as composition is concerned. What is so valuable regarding these objects is that they regularly intersect the Earth's atmosphere, and a number of them fall to the surface of the Earth where they can be collected and examined. Apart from a small amount of lunar material, this is the only source of cosmic material that is available for laboratory investigation. Any conclusion one can reach regarding this material could therefore be of immense value.

Unfortunately, even though some of the facts are well established, different authors give very different interpretations of their meaning. One reason for this divergence is obvious: the meteorite sample generally available represents only a very minor fragment (about 10^{-3}) of the original parent body, and in such a large-scale extrapolation, personal bias on the part of the extrapolator inevitably plays a part.

Most of the evidence gathered comes from the study of chondrites, which are the most abundant of the stony meteorites. These chondrites contain chondrules, millimetre-sized silicate spheres that have the appearance of frozen droplets from some molten substance. The composition of the chondrules is basically olivine [$(Mg, Fe)_2SiO_4$], pyroxine [$(Mg, Fe)SiO_3$] and plagioclase feldspar [solid solution of $CaAl_2Si_2O_8$ and $NaAlSi_3O_8$]. In some chondrites, glass is found in place of the feldspar. These chondrules are embedded in a matrix which is somewhat more finely grained. Millimetre-sized metal particles (nickel–iron) or troilite (FeS) particles are also present.

There are five chondrite classes that are usually recognized. Enstatite chondrites are highly reduced, which means that they contain iron only as metal and some troilite. Carbonaceous chondrites are highly oxidized. In between are three classes for which the oxidization states and total iron content are intermediate—the H, L and LL chondrites. Collectively, these three classes are called the ordinary chondrites.

During the last decade it has become apparent that most of the properties of chondrites were established prior to the accretion (if any) of their parent body, which might have been asteroidal. These chondrites, therefore, contain a very accurate record of conditions at a very early stage in the evolution of the solar system. The problem is to decipher this record—firstly by translating the structure, composition and mineralogy which have been found into temperature, pressure, and chemical surroundings, and then by comparing these conditions with suggested conditions arising from a theory for the origin of the planets.

There have been many review papers dealing with accumulating information on meteorites and its interpretation (e.g. Anders, 1971, 1972). Since this is outside the main theme of this work, I shall only briefly summarize the results here, paying particular attention only to any restrictions which this work places on the type of theory that is allowable, or to the stages which must have existed in the pre-planetary system if chondrites of the type found are to be generated. The following (taken from Anders, 1972) would appear to cover the generally agreed points.

(1) Material rich in Al, Ca, Ti and so on went its separate way in the inner solar system, the Moon and the Earth being rich in such elements, but the ordinary and enstatite chondrites being depleted. Some mechanism must be

present to allow for this, the obvious one being that these materials condensed early.
(2) A metal-silicate fractionation has affected several meteorite classes, as well as the inner planets, and must therefore reflect some widespread occurrence.
(3) In the source region of the meteorites, part of the primary condensate was remelted and outgassed by some process.
(4) Material in the neighbourhood of the asteroidal belt (if this forms the meteorite source) must have been at a temperature of 450 K with a pressure of the order of 10^{-5} atm at some stage.

Rather than attempt to form theories for the origin of the planets from the above list, we shall use it to check the acceptability of proposed theories.

1.9. The Main Features of the Planetary System

In the preceding sections, we have considered relevant information regarding individual members of the solar system. In this section, we shall try to bring them all together and produce a summary of what the main features of the solar system are, bearing in mind Section (1.1) which states that what we should really be doing is categorizing the properties of planetary systems in general, but that we cannot do this because of lack of information. These main features should be explained, or should occur as a direct consequence of a postulate, in all acceptable theories for the formation of the planets. These main properties are as follows:

(1) There exists a central condensation, the Sun, which is many times (a factor of 750) more massive than the sum of the remaining parts of the system.
(2) The Sun rotates very slowly, both in relation to the angular momentum present in interstellar gas clouds and in relation to the angular momentum of the planets. In fact, the sum of the angular momenta of the planets about

the Sun is about 200 times larger than that of the Sun about its own axis. It is, of course, almost inevitable that planets in orbit about the Sun will have a greater angular momentum than the Sun, for they are, in effect, held in orbit by a centrifugal force balancing gravity whereas for the Sun centrifugal force must be less than gravity. What is wanted is an explanation of how the planets came to be in orbits at such large distances from the Sun.

(3) There are nine known planets in orbit about the Sun. We are not, therefore, looking for a mechanism which can form binary stars with very unequal masses, but rather a mechanism which is capable of forming many objects about a central body.

(4) The orbits of the planets all lie close to a well-defined plane, so that the planetary system is essentially two-dimensional. The rotation of the Sun about its own axis is also essentially in this plane. Care must be taken on this point since the evidence from Barnard's star (Black & Suffolk, 1973) suggests that it is a system which does not possess this property. It may be that the alignment process for planets is not a completely efficient process, so that exceptions can exist.

(5) All the planets move in the same, prograde, sense round their orbits. There is also a tendency for the planets to rotate about their own axis in the same sense; while the majority of the satellites also have prograde orbits.

(6) There exists a clear division in the chemical composition of the planets, which corresponds both to their different spatial position and to their different masses. This has been tabulated in Table 1.4.

(7) The orbital distances of the planets are roughly given by the Titius–Bode law. This cannot be considered independently of the angular momentum of the planets. Nevertheless, some explanation is needed for the consistent increase in the distance between the planets (or in their angular momentum per unit mass) as one moves away from the Sun.

(8) There are also a number of smaller objects to be found

in the present day system. The distribution of mass between the various bodies is given in Figure 1.2.

Later, we shall be describing some of the theories for the formation of the planets that have been proposed. Their success should be judged first in the light of their ability to explain the above general features. If they can satisfy these, then there are the more detailed points, outlined in the preceding work, which also need to be satisfied.

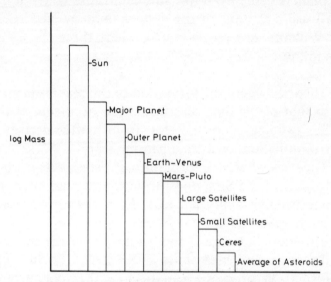

Figure 1.2   Mass distribution in the solar system.

1.10. Classification of Theories

It is, of course, possible to present the theories that have been proposed in chronological order starting with the oldest and finishing with the newest. Since there are many different kinds of theories, starting with different assumptions and making use of different physical effects, no clear picture emerges from following the historical sequence. We shall, therefore, divide the theories into categories, so that theories in any one category all have something in common with one another. In this way it becomes possible to see clearly the development of certain ideas and to see that there are only a few central themes in all the theories.

There are many ways one might choose to classify theories, an obvious one being into two classes, those which the author likes and those which he does not. Williams (1974) suggested two categories, depending on whether the material in orbit about the Sun from which the planets are to form was in a continuous form, or whether it was discrete. Much earlier, Jeans (1931) had also devised a two-class system: those theories where the material for the planets came from the Sun, and those where it did not. We shall use a variant of the system used by McCrea (1963) and by Williams & Cremin (1968), where three classes are defined as follows.

(1) Theories where the formation of the planets is unrelated to that of the Sun, and formation occurs, after the Sun has become a normal star, from matter derived either from the Sun, or from a passing star.
(2) Theories where the formation of the planets is unrelated to that of the Sun, and formation occurs, after the Sun has become a normal star, from matter derived from interstellar space.
(3) Theories that regard the formation of the planets as a direct consequence of the formation of the Sun. The two formation processes can occur either concurrently or consecutively.

We shall discuss theories in each of these categories in turn, at the same time mentioning briefly the shortcomings and weak points of the particular theory being discussed. Actual comparison of theories will be left until Chapter 5.

2. The Tidal and Other Related Theories

2.1. Introduction

We shall now discuss theories in the first category defined in Chapter 1: namely theories according to which the planets formed after the Sun had become a normal star, and where the material from which the planets are to form originated either in the Sun or in another nearby star. These type of theories have become known as *Tidal Theories*, basically because the main mechanism envisaged to remove material from the Sun, or star, is the tidal action of the passing star, or Sun, respectively. One of the first theories of this type was proposed by Buffon in 1745. He suggested that the planets condensed out of material ejected from the Sun as a consequence of a comet colliding with it. In view of the information given earlier about comets, Buffon's theory is fantastic, but it should be borne in mind that he had no way of knowing the mass of a comet, and if it is assumed that a comet has a density similar to the Sun (or the planets) then the estimate for the mass turns out to be much in excess of that of the Earth. Indeed, Buffon had already estimated the mass of the great comet of 1680 as 28 000 times that of the Earth or nearly a tenth that of the Sun. Given such views regarding comets, Buffon's supposition regarding the effects of the collision is not that incredible. By 1880, the true nature of comets was known and so the clear impossibility of Buffon's theory became apparent. It was modified by Bickerton (1878), who replaced the comet–Sun collision with a collision between the Sun and a passing star. This collision was assumed to cause a nova-like outburst, and the planets formed by direct condensation from this emitted material.

In this century, Arrhenius (1913) considered the head-on collision of two stars which, according to his hypothesis at least, results in one star and a gaseous filament. See (1910) considered the collision of two nebulae. Presumably he had in mind a gas-cloud type of nebula (remember that this was before the great debate regarding the nature of nebulae). He conjectured that this would generate one proto-Sun which would subsequently capture the planets from the other remnants of the collision.

Slightly earlier a variation on this theme, whereby a stellar encounter led to a remnant from which the planets formed by direct condensation, was proposed independently by Chamberlain (1901; 1927) and Moulton (1905). They considered the encounter of the Sun and another star in which the field of the star served to intensify the solar activity on the side of the Sun facing it. This resulted in the ejection of great clouds of gaseous material from the Sun. These clouds were also influenced by the passing star in that they were given a transverse component to their motion, thus accounting for the large angular momentum of the planets.

As these clouds cooled, liquid drops which later solidified were assumed to condense out of them. The planets were formed by the accumulation of these solid drops, while the residual matter acted as a resisting medium that would round off the orbits of the growing planets and so generate the present features of the planetary system. These small, solid particles were called planetesimals, and this terminology has now become universally accepted. Chamberlain and Moulton were the first two authors to suggest that the planets could have formed by the agglomeration of small, solid particles rather than from the direct condensation of a hot fluid, and so, indirectly at least, led the way to many a modern theory that makes use of such ideas in order to explain the chemical distribution in the system. These ideas were, however, criticized by Jeffreys (1924), who argued that the planetesimals would collide frequently, and with such a high relative velocity, that they would vaporize rather than accrete into planets.

2.2. The Tidal Theories of Jeans and Jeffreys

It is, of course, with the names of Jeans and Jeffreys that the tidal theories are most commonly associated. In a series of papers, Jeans (1916; 1917a, b) investigated in some detail the interaction between two stars. The first paper (1916) showed that an actual collision was not necessary in order to remove material from one of the stars. In a close encounter, the gravitational interaction would cause a bulge to form on one of the stars, from which material could be removed. Let the tide-generating body be of mass M, and, for simplicity, let us assume that it is constrained to remain spherical throughout. Take coordinates (r, θ) at the centre of the other mass, with $\theta = 0$ along the line of centres of the two masses (Figure 2.1), and let the separation of the centres be R. The potential at a point with coordinates (r, θ) is

$$\frac{GM}{(R^2 + r^2 - 2Rr \cos \theta)^{1/2}} = \frac{GM}{R} + \frac{GMr \cos \theta}{R^2}$$
$$+ \frac{GMr^2 P_2(\cos \theta)}{R^3} \qquad (2.1)$$

where $P_n(\cos \theta)$ are the Legendre polynomials. The term GM/R is a constant, and so produces no tidal force (remember that potential is the integral of a force). The term $(GMr \cos \theta)/R^2$ gives rise to the uniform force field GM/R^2, which is simply the mutual gravitational attraction of the centres of mass, and can be removed by selecting a suitable moving frame of reference. The first term that generates a tide in the above expansion is

$$\frac{GMr^2 P_2(\cos \theta)}{R^3}. \qquad (2.2)$$

If we assume that this is the only term of importance, then in cartesian coordinates with x along the $\theta = 0$ axis, this potential becomes

$$\mu(x^2 - \tfrac{1}{2}y^2 - \tfrac{1}{2}z^2), \qquad (2.3)$$

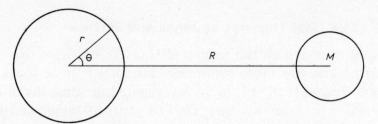

Figure 2.1. Geometric configuration for calculating tidal effects.

where $\mu = GM/R^3$. Jeans points out that this potential is very similar to that obtained for a rotating mass.

If the potential at any point of the tide-experiencing body is V, then the equilibrium of this body requires that for every point on its surface

$$V + \mu(x^2 - \tfrac{1}{2}y^2 - \tfrac{1}{2}z^2) = \text{Constant}. \qquad (2.4)$$

From this equation it is clear that ellipsoidal configurations exist. If we assume that a possible equilibrium figure is given by

$$\frac{x^2}{a^2} + \frac{y^2}{b^2} + \frac{z^2}{c^2} = 1,$$

then its potential is

$$V = -\pi\rho abc \int_0^\infty \left(\frac{x^2}{a^2 + \lambda} + \frac{y^2}{b^2 + \lambda} + \frac{z^2}{c^2 + \lambda} - 1 \right) \frac{d\lambda}{\Delta}, \qquad (2.5)$$

where $\Delta^2 = (a^2 + \lambda)(b^2 + \lambda)(c^2 + \lambda)$. The left-hand side of equation (2.4) is therefore a linear function of x^2, y^2 and z^2. By substitution and equating coefficients, we find that the ellipsoidal shape is a possible equilibrium configuration, provided that

$$\frac{2}{abc}\left(\frac{1}{b^2} - \frac{1}{c^2}\right) = \left(\frac{1}{a^2} + \frac{1}{b^2} + \frac{1}{c^2}\right) \int_0^\infty \frac{(c^2 - b^2)\,d\lambda}{(b^2 + \lambda)(c^2 + \lambda)\Delta}. \qquad (2.6)$$

This obviously reduces directly to the two equations

$$b^2 = c^2, \qquad (2.7)$$

$$\frac{2}{ab^3c^3} = \left(\frac{1}{a^2} + \frac{1}{b^2} + \frac{1}{c^2}\right) \int_0^\infty \frac{d\lambda}{(b^2 + \lambda)(c^2 + \lambda)\Delta}, \quad (2.8)$$

which allow two series, one of spheroids and one of ellipsoids. For the ellipsoidal series, Jeans found that equation (2.4) reduces to

$$\frac{\mu}{2\pi\rho abc} = -\int_0^\infty \frac{\lambda\, d\lambda}{(b^2 + \lambda)(c + \lambda)\Delta}, \quad (2.9)$$

requiring a negative value for μ, which is impossible. Therefore only the spheroidal series needs to be considered. Jeans found numerically that if e is the eccentricity of the spheroid, then as e increases so does μ until the values

$$e = 0 \cdot 882579$$

$$\frac{\mu}{\pi\rho} = 0 \cdot 125504$$

are reached, after which μ steadily decreases. (The lengths of the axes for this critical value of e are approximately in the ratio of $17:8:8$.) This means that all spheroids with eccentricity greater than $0 \cdot 882579$ are unstable, and Jeans concluded that mass loss would occur. Jeans, in fact, investigated many cases much more complex than the one we have described. He considered such things as the zonal harmonics of the surface, in which case the limiting value for μ becomes

$$\frac{\mu}{\pi\rho} = 0 \cdot 109131$$

rather than that given above. He also included terms other than the P_2 term in the tide-generating potential, as well as considering the fact that the tide-raising body had a finite size. In all cases, he concluded that mass loss would occur eventually.

Jeans next considered the stability of a stream of fluid ejected from one of the stars. He assumed the usual equation of motion

$$\ddot{\mathbf{r}} = \mathbf{F} - \frac{1}{\rho}\mathrm{grad}\, P, \quad (2.10)$$

where $\mathbf{F} = (X, Y, Z)$ represents all the forces acting, and P is the pressure. He also assumed that P is a function of ρ only (the isothermal approximation), so that

$$\int \frac{dP}{\rho} = \phi(\rho). \tag{2.11}$$

Equation (2.10) therefore becomes

$$\ddot{\mathbf{r}} = \mathbf{F} - \operatorname{grad} \phi. \tag{2.12}$$

He then considered the stability of this equation by taking $\mathbf{r} = \mathbf{r}_0 + \delta \mathbf{r}, \mathbf{F} = \mathbf{F}_0 + \delta \mathbf{F}, \rho = \rho_0 + \delta \rho$. This gives an equation whose x component is

$$f_x - \frac{\partial f_x}{\partial x}\xi - \frac{\partial f_x}{\partial y}\eta - \frac{\partial f_x}{\partial z}\zeta + \frac{\partial^2 \xi}{\partial t^2} = X_0 + \delta X - \frac{\partial}{\partial x}\phi(\rho + \delta\rho) \tag{2.13}$$

where $f_x = \ddot{x}$, and (ξ, η, ζ) are the components of $\delta \mathbf{r}$. However, x is a solution of the equation of motion, and so

$$f_x = X_0 - \frac{\partial \phi(\rho)}{\partial x},$$

which leads to the equation

$$\frac{\partial^2 \xi}{\partial t^2} - \frac{\partial f_x}{\partial x}\xi - \frac{\partial f_x}{\partial y}\eta - \frac{\partial f_x}{\partial z}\zeta = \delta X - \frac{\partial}{\partial x}\left[\frac{\partial \phi}{\partial \rho}\delta\rho\right]. \tag{2.14}$$

If we average over the whole stream, then

$$\overline{\frac{\partial f_x}{\partial x}} = \overline{\frac{\partial f_x}{\partial y}} = \overline{\frac{\partial f_x}{\partial z}} = 0,$$

and so the equation of motion of the perturbation is

$$\frac{\partial^2 \xi}{\partial t^2} = \delta X - \frac{\partial}{\partial x}\left[\frac{\partial \phi}{\partial \rho}\delta\rho\right], \tag{2.15}$$

with similar equations for each of the other two components. Differentiating each of these three component equations with respect to x, y and z and combining gives

$$\frac{\partial^2}{\partial t^2}\text{div}(\delta \mathbf{r}) = \text{div}\,\delta \mathbf{F} - \nabla^2\left[\frac{\partial \phi}{\partial \rho}\delta\rho\right]. \tag{2.16}$$

Poisson's equation gives

$$\text{div}\,\mathbf{F} = -4\pi G\rho \quad \text{so that} \quad \text{div}(\delta\mathbf{F}) = -4\pi G\delta\rho.$$

Also, from equation (2.11),

$$\frac{\partial \phi}{\partial \rho} = \frac{1}{\rho}\frac{\partial P}{\partial \rho},$$

so that equation (2.16) becomes

$$\frac{\partial^2}{\partial t^2}\text{div}(\delta \mathbf{r}) = -4\pi G\delta\rho - \nabla^2\left(\frac{\partial P}{\partial \rho}\frac{\delta\rho}{\rho}\right). \tag{2.17}$$

Denoting $\delta\rho/\rho$ by S, we have $\text{div}\,\delta\mathbf{r} = -S$, and equation (2.17) gives

$$\frac{\partial^2 S}{\partial t^2} = 4\pi G\rho S + \nabla^2\left(S\frac{\partial P}{\partial \rho}\right). \tag{2.18}$$

If, for simplicity $\partial P/\partial \rho$ is taken to be nearly uniform, then equation (2.18) has a solution of the form

$$S = S_0 \exp\{i(qt - 2\pi x/\lambda)\}, \tag{2.19}$$

which represents the propogation of waves of frequency $q/2\pi$, and wavelength λ. Substitution of (2.19) into equation (2.18) yields

$$q^2 = \left(\frac{2\pi}{\lambda}\right)^2\frac{\partial P}{\partial \rho} - 4\pi G\rho.$$

The nature of the solution changes to that of an exponentially growing disturbance when

$$4\pi G\rho = \left(\frac{2\pi}{\lambda}\right)^2\frac{\partial P}{\partial \rho},$$

that is when

$$\lambda = \sqrt{\frac{\pi}{G\rho} \frac{\partial P}{\partial \rho}}. \qquad (2.20)$$

This shows that the stream will break up into blobs each of length λ given by equation (2.20), the blobs rapidly condensing to high densities.

At that time, Jeans's theory was very attractive because he claimed to have shown mathematically that one star could tidally influence another, leading to the loss of matter. He had further shown that this ejected matter would break up into discrete blobs which would rapidly condense, presumably to planetary densities. Such ideas for the formation of the planets were given a further boost when Jeffreys (1918) showed that it was not possible to obtain the existing planetary system, with its present mass and angular momentum distribution, from the slow contraction of a single gaseous mass moving initially in any plausible way, whatever initial density distribution was allowed. Of course, Jeffrey's argument did not allow for the redistribution of angular momentum within the gaseous condensation, either because of viscous effects or because of electromagnetic phenomena, and the gas cloud under consideration was taken to be one single entity throughout. It is primarily the realization that such effects can be important that has led to the resurgence of nebular theories for the origin of the planets. At the time, however, Jeffrey's conclusion appeared to dispose of virtually all theories with the exception of those of the tidal type, especially since Jeans (1917a) had already disposed of the possibility that the system came about as a result of the fission of a single rotating body contracting in isolation, angular momentum again being the over-riding criterion.

As mentioned earlier, Jeans had shown that the effect of a tidal dissipation was in many ways similar to that of rotation. In 1928, he accordingly modified his tide-generating potential to include the effects of rotation. The equation (2.4) therefore now contains on the left-hand side the additional term

$$+\tfrac{1}{2}\omega^2(x^2 + y^2). \qquad (2.21)$$

It is fairly obvious that the gravitational part can be somewhat smaller and yet still lead to mass loss, so the phenomenon becomes somewhat easier to achieve. But the symmetry has now been lost, and Jeans found that mass loss would in fact occur from an ellipsoidal shape in the form of two filaments, one protruding from each end of the longest axis. These filaments are similar to the one-dimensional stream which he investigated and found to break up into droplets. In this work, he showed further that if the density of the filament had a suitable value, then these droplets would be of planetary mass (that is of the order of 10^{26} kg). From a further investigation into the ejection of the filament, Jeans (1931) concluded that it would be cigar-shaped. Since the density is constant, this meant that there would be more mass near the centre of the filament; and since the instabilities fragmented the filament into equal-length drops, this, in turn, implies that the central planets (namely Jupiter and Saturn) would be more massive than the other planets.

One of the difficulties of the tidal theory as outlined by Jeans was to explain the individual rotation of the planets, and this difficulty was recognized by Jeffreys (1929). He therefore revived, and suitably modified, Buffon's original idea of a collision with the Sun. In the modified version, a passing star grazed the Sun. This collision was claimed to tear out a cigar-shaped filament very similar to the one produced by rotation and tidal effects, hence apparently giving the correct mass distribution for the planets. It was further claimed that viscous effects would set up a rotation of the correct order of magnitude in each of the resulting drops.

The concept of the tidal theory as envisaged by Jeans and Jeffreys is essentially very simple, and is in many ways attractive because of this simplicity. (The mathematics used to justify it is, of course, much more complicated, but that is a different matter). Because of the action of a passing star (either a collision or a tidal effect), material is drawn out from the Sun in the form of a cigar-shaped filament. This filament lies in the plane defined by the relative motion of the two stars, and all subsequent motion of the filament and its fragments will be in the same direction in this plane. This filament, because of instabilities in it, breaks up into a number of drops, each of which contracts to become a planet. Because of the initial shape of the filament, the more massive planets will be found towards the centre of the system of planets, where Jupiter and Saturn are to be found. Residual material from the ejected matter was claimed to be responsible for rounding the orbits into their present circular form; while the viscous effects within the filaments explained the individual rotation of the planets. It was even claimed that the partial break-up of a planet when it first returned to the vicinity of the Sun, before the orbits were rounded off, could explain the satellite systems. Of course, nothing was then known regarding the chemical composition of the planets, and so this caused no difficulty.

In spite of this rather impressive list of explanations, some individuals were soon offering criticism—in a qualitative form by Luyten (1933) and quantitatively by Nolke (1932), Russell (1935) and Spitzer (1939). We shall next discuss the objections to the tidal theories proposed above.

2.3. Objections to the Tidal Theories

Nolke (1932) discussed the tidal dissipation of the planetary mass. If M_P denotes the mass and R_P the average radius of the material removed from the Sun (which is, itself, of mass $M_\odot$ and radius $R_\odot$), and if the centres are separated by a distance d, then there is a difference in the gravitational attraction due to the Sun between the centre of the planet and its face nearest to the Sun of

$$\frac{GM_\odot}{(d-R_P)^2} - \frac{GM_\odot}{d^2} \qquad (2.22)$$

If the material is not to be disrupted, this shearing force must be balanced by the material's own gravitational field, namely

$$\frac{GM_P}{R_P^2}.$$

We therefore require

$$\frac{M_\odot}{(d-R_P)^2} - \frac{M_\odot}{d^2} < \frac{M_P}{R_P^2}, \qquad (2.23)$$

which on tidying up becomes

$$\frac{M_P}{M_\odot} > \frac{R_P^3}{d^2} \frac{(2d-R_P)}{(d-R_P)^2}. \qquad (2.24)$$

Now, since the matter is torn from the Sun, the initial value for d is $R_P + R_\odot$; so that inequality (2.24) becomes

$$\frac{M_P}{M_\odot} > \frac{R_P^3(R_P + 2R_\odot)}{(R_P + R_\odot)^2 R_\odot^2}. \qquad (2.25)$$

If $R_P \geqslant R_\odot$, the right-hand side of the above is of order unity, which means that M_P, the mass of the filament, must be about the same as the solar mass, in contradiction to the hypothesis implied in the tidal theory. We need therefore only consider the case when $R_P \ll R_\odot$, for which inequality (2.25) becomes

$$\frac{M_P}{M_\odot} > \frac{2R_P^3}{R_\odot^3}. \qquad (2.26)$$

If we denote the mean densities of the planetary and solar material by ρ_P and $\rho_\odot$ respectively, then the right-hand side of the inequality becomes $2M_P \rho_\odot / M_\odot \rho_P$ and the whole inequality reduces to

$$\rho_P > 2\rho_\odot.$$

This is also an impossibility since the denser central regions of the star will remain within the star. Nolke therefore concluded that the proposed tidal theory was not tenable.

The next objection came on a point where the tidal theories appeared to be strongest, namely the angular momentum of the planets. Stated in its simplest form, the objection is that it is impossible to transfer angular momentum to any material without also transferring energy, and that if the present angular momentum of the planets is given to the material, then it also receives an amount of energy which is greater than is required for escape from the system. Conversely, if the energy is correctly selected, then the angular momentum is small, and the material's orbit intersects the surface of the Sun. Russell (1935) carried out some order-of-magnitude calculations, and concluded that planets could not be formed at a distance further out than about four solar radii (about a twentieth of the orbital distance of Mercury). Much later, Lyttleton (1960) came to much the same conclusion having investigated the restricted three-body problem with the aid of a computer. The orbits of the small third body either intersected the surface of one of the primaries or went to infinity, though this is not the case if magnetic effects are included (Sarvajna, 1970). By the time Lyttleton's work was published, the tidal theories had been superseded by other theories.

Nolke (1930) and Russell (1935) had also questioned the idea that residual material could be effective in rounding off the highly-elliptical orbits which the tidal theory predicts for the planets. This is a very difficult question to tackle, since the answer obviously depends on the amount of material that is thought to be left over, the manner in which this material is moving, and the density of the planet whose orbit is being rounded. For the tidal theories, where most of the material goes into planets, the planets are fairly dense and the orbits very elongated, rounding off is unlikely to occur. For nebular theories, where the orbits are more newly circular and there is much more gas about, the answer is not so obvious.

Another objection came from Spitzer (1939), who was concerned with the stability of the material torn from the Sun. His approach was concerned only with the internal energy which any material removed from a star must obviously possess. To simplify the calculations, let us assume that planetary material is a spherical condensation with mass M_P, radius R_P, and density ρ_P. Its self-gravitational energy is therefore given by

$$U = -\frac{C_1 G M_P^2}{R_P}, \qquad (2.27)$$

where C_1 is a constant whose exact value depends on how centrally-condensed the planetary material is: in reality, its value is always close to, but less than, unity. If the material is at a temperature T_P, then its internal energy is given by

$$\frac{3}{2}\frac{\mathscr{R}}{\mu}T_P M_P, \qquad (2.28)$$

where $\mathscr{R}$ is the gas constant and μ the mean molecular weight. The filament will stay as a bound system only if its total energy is negative, that is if

$$\frac{3}{2}\frac{\mathscr{R}}{\mu}T_P M_P < \frac{C_1 G M_P^2}{R_P}. \qquad (2.29)$$

Eliminating the radius R_P, in favour of the mean density ρ_P, gives the condition for existing as a bound state as

$$\frac{3}{2}\left(\frac{3}{4\pi}\right)\frac{\mathscr{R}T_P M_P^{-2/3}\rho_P^{-1/3}}{\mu C_1 G} < 1. \qquad (2.30)$$

Now, if the material has a planetary mass, that is 10^{27} kg, then it must have been extracted from the deep solar interior, so that $T_P \sim 10^6$ K and $\rho_P \sim 10^3$ kg m^{-3}. Inserting these numerical values, together with those for the universal constants, shows that the left-hand side of inequality (2.30) has a value of order ten, which means that the inequality is not satisfied. Consequently the material has positive total energy and will disperse rather than condense. The only way in which the material can avoid this outcome is to cool sufficiently rapidly that it becomes stable before it has time to disperse. Kramers & Burgers (1946) have shown that the expansion of the outer layer of a dispersing filament is essentially at the speed of sound, namely with a velocity

$$v \sim \sqrt{\frac{\mathscr{R}T_P}{\mu}}. \qquad (2.31)$$

The filament will therefore double its dimensions in a time of order τ_1, where

$$\tau_1 = R_P/v. \qquad (2.32)$$

Inserting the adopted numerical values into this equation leads to a time of the order of a few hours. Dispersal is therefore very rapid. The fastest cooling rate is the black-body rate, namely an energy loss rate

$$L \sim \pi a c R_P^2 T_P^4, \tag{2.33}$$

where a is the radiation constant and c the velocity of light. An estimate of the time taken to reduce a filament's internal energy, W, by half is given by τ_2, where

$$\tau_2 = W/L. \tag{2.34}$$

What is, in fact, of more interest is the ratio of the cooling rate to the expansion rate, namely τ_2/τ_1. With the adopted numerical values, it transpires that $\tau_2/\tau_1 \sim 10$, which means that the expansion takes place much quicker than the cooling; consequently dispersal must take place. The analysis given above is very simplified and contains many assumptions. Spitzer (1939) carried out a much more careful and detailed investigation, but his conclusion was similar to that reached above. Consequently, planets cannot condense from material torn out of a normal star.

These are the major criticisms of the tidal theories. There is an additional modern criticism, which is that no attempt is made to obtain the correct chemical composition of the planets. Indeed, the reverse is true: if the correct mass is obtained (as the theory claims) with no attempt at a chemical segregation of the material, then all planets will have the same composition. However, this criticism only came to light after the theory in its original form had been abandoned, and so does not feature in the literature.

The tidal theories were modified in view of the above objections, though by no means all the questions that had been raised were answered. However, we shall next discuss such modifications as were made.

2.4. Theories Involving Multiple Stars

In 1936, a modification to the theory was proposed whereby a binary system played an important part in the development of the planetary filament. This was primarily an attempt to overcome the objection of Russell (1935) that planets could form only as far out as four solar radii on the traditional tidal theory. The suggestion was made by Lyttleton (1936; 1938a, b) that the Sun once formed part of a binary system. The companion star then either underwent a collision or a close encounter with a third star that happened to pass the system. This encounter resulted in the ejection of a gaseous filament from the companion star which was subsequently captured by the Sun. It also led to the companion star being detached from the Sun, which, of course, explains why no companion star is present today. In this way Nolke's objection regarding the tidal disruption of the planetary filament could also be overcome, since the filament never went very close to the Sun, and the companion star and the passing star were both removed from the vicinity before they could cause any such disruption. Spitzer's objection, of course, remains, though it had not been formulated at the time Lyttleton proposed the above modification. Lyttleton (1937) also pointed out that with the assumed density of the filament, the terrestrial planets were not massive enough to be able to condense out of it. In consequence, he postulated that originally one massive protoplanet was formed. This became rotationally unstable as it contracted and fragmented into two major components, Jupiter and Saturn, together with a connecting filament from which the remainder of the planets condensed, being able to do so because the gravitational fields of Jupiter and Saturn in part compensated for the field of the Sun.

One aesthetic objection to the tidal theories, which had been voiced by many authors, was that a suitable encounter between the Sun and another star was extremely unlikely. In an attempt to overcome this criticism, Lyttleton (1940; 1941a, b) further modified the theory. He suggested that the Sun was originally part of a triple system, consisting of the Sun and a close binary pair. In common with all other stars, this system would accrete interstellar matter, and because such matter would have low angular momentum relative to the binary pair, this accretion would decrease their angular momentum per unit mass, resulting in a decrease in their separation. Eventually the two stars would come into contact with one another and would become, temporarily at least, a single star. Such a single star would be rotationally unstable, and so fission into two parts, with a mass ratio different from that of the original pair, would occur. After this fission, both main components would escape from the Sun, but a filament that had been formed between them would remain. The planets then formed (in the manner previously described) from this filament.

Gunn (1932) had also realized how unlikely a stellar encounter of the type envisaged by the tidal theorists was. His remedy was to revive the type of theory where the Sun fragmented because of instabilities. As previously mentioned, Jeans (1917a) had shown that a simple fission process would not work. Gunn therefore evoked electromagnetic means to induce the instability which caused the fission. The planets formed in the usual way from a filament drawn out between the stars. The objections of both Spitzer and Nolke apply to this theory, but it is ingenious enough to deserve a mention as a short diversion from the main theme.

In an attempt to overcome the objection of Spitzer as well as that of Nolke, Dauvillier (1942a, b) assumed that the filament would be ionized (as indeed it could be if the temperature were high enough for Spitzer's objection to be valid). He showed that the electromagnetic effects which followed from the assumption that the filament was ionized could stabilize the filament long enough to allow condensation into planetary twins to take place. He further showed that condensations would occur at exponentially increasing distances from the Sun (an approximation to the Titius–Bode law). Usually the twin planets merged into one while still in a gaseous state, but Dauvillier thought the Earth–Moon system may have been an exception. This work and that of Lyttleton showed great ingenuity, and introduced many new, and by no means invalid, concepts; but it is interesting to note that the attempts were still intended to patch up a crumbling theory, rather than to strike out with completely new ideas.

The last major development of tidal theories that was to occur for some time was proposed by Hoyle (1944). He considered the Sun to have been originally a part of a binary system.

However, he suggested that the companion star was not involved in any encounters, nor did it become rotationally instable. Rather, it went through its normal evolutionary sequence and became a nova. Material ejected in the explosion of the latter was captured by the Sun, and the planets formed from this material. In order to generate a planetary system with the properties of the existing system, he found that the amount of matter ejected must have been about $M_\odot/10$. This is far too much mass to have been ejected from a nova, and so he modified the theory (Hoyle, 1945) by replacing the nova outburst with a supernova explosion. This modification had the added advantage of removing the unwanted stellar remnant. Values for the various parameters involved were taken from observations of the Crab Nebula, and it was found that all the requirements of his theory were satisfied. He also agreed with Lyttleton about the formation of the terrestrial planets, and proposed (Hoyle, 1946b) that they were formed from the rotational break-up of a protoplanet.

No further development of the tidal theories was to occur until a new theory was proposed by Woolfson (1960). This theory and its further developments will be discussed in Section (2.6). But we first mention an interesting theory that hardly fits into the main stream of development of the tidal theory.

2.5. An Unusual Theory

The theory of Banerji & Srivastava (1963) considered a spherical magnetic star of mass $9M_\odot$, which is oscillating radially with a small amplitude. They showed that the nearby passage of a star of similar mass increases the amplitude of the oscillation of the magnetic star, rendering it unstable. The instability leads to the ejection of mass from the star, and the planets form from this. Banerji & Srivastava concluded that the two stars need not pass very close to one another in order to cause the instability to develop; nor does the relative velocity have to be high in order to produce the required angular momenta of the planets. Even though this theory has been presented in terms of the formation of the planets, it is really a theory about the formation of a stellar envelope from which planets might develop. If there were not other simpler ways of generating such an envelope, or solar nebula, perhaps this theory would receive closer study.

2.6. Woolfson's Theory of Planetary Formation

In some ways, it is unfair to discuss the theory proposed by Woolfson (1960) as a sequel to other tidal theories, since it is a new theory rather than a modification of previous theories. However, obtaining the planetary material does depend on the tidal effect of a star (the Sun), and so it seems appropriate to discuss the theory in this chapter. The event takes place at the epoch when a stellar cluster is just forming. Some stars (including the Sun) will have evolved through their pre-main-sequence phase and become normal stars, while others will still be at an early stage of contraction. Williams & Cremin (1969) investigated the ages of stars in the young cluster NGC 2264, finding that a considerable spread in the formation times of stars existed, and also that stars of about one solar mass tended to form first.

At this epoch in the development of a cluster, the stars are generally thought to be closer together than at present, and Woolfson makes this point in order to overcome the aesthetic objection regarding the improbability of stellar encounters. It is generally thought that widely-separated binaries formed as a consequence of such encounters during this epoch. The most detailed version of Woolfson's theory was published in 1964: in this, he considers the encounter between the Sun and a protostar of mass 3×10^{29} kg (about 1/7 of a solar mass). The closest approach distance is taken to be $6 \cdot 67 \times 10^{12}$ m, comparable to the dimensions of the planetary system. Woolfson takes the protostar to have a mean radius of the order of 3×10^{12} m, so that its mean density and mean temperature are both very low, namely 4×10^{-9} kg m^{-3} and 30 K respectively. By having such a low value for the temperature, Spitzer's objection is made invalid, while the low density allows tidal distortion to occur at the large distance from the Sun which is envisaged. This obviously overcomes Russell's objection, since the ejected matter now finds itself initially at large distances from the Sun.

Woolfson does not analyse the total effect by complex mathematics in the same manner as Jeans. Instead, he simulates the encounter using a computer, though he does show that the ejection of the material is possible by reference to the earlier work of Jeans. It is assumed that the Sun moves past the star which is to be distorted, so that the potential is given by adding the contribution from the distorted star to that given by equation (2.1). As before, the potential arising from the mutual attraction of the centres of gravity is removed, cancelled by a component due to the relative motion of the stars. The equipotential surfaces are therefore given by

$$\frac{M}{r} + \frac{M_\odot}{(r^2 + R^2 - 2rR\cos\theta)^{1/2}} - \frac{M_\odot r \cos\theta}{R^2} = \text{Constant} \quad (2.35)$$

where $M_\odot$ is the mass of the Sun, and M the mass of the distorted star. This can also be expressed as

$$\frac{M}{r} + \frac{M_\odot}{r'} - \frac{M_\odot x}{R^2} = \text{Constant},$$

where r' and x are as shown in Figure 2.2. Also included in this figure are a few members of the family of equipotentials given by this equation. If the dimensions of the star exceed that of the critical equipotential (heavy line), then mass loss occurs. It should be noted that the above is standard work for discussing mass exchange in binary stars, and is therefore presumably above suspicion.

Woolfson produced a computer model of such a distorted star, where its interior is represented by a series of discrete point masses. The model considered is two-dimensional, and most of the mass is concentrated in one point at the centre. The remainder of the star is represented by a network of points in the outer annular region, and mutual gravitational attraction between these points is considered. In a real star, pressure would

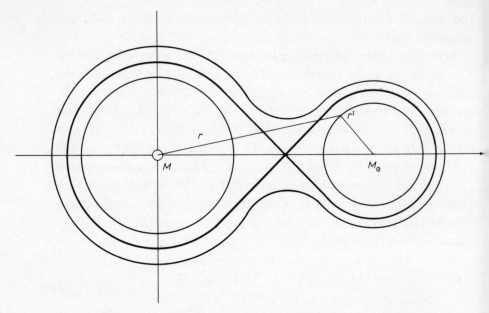

Figure 2.2 Equipotential surfaces for the Sun and a distorted star.

keep such points apart, and so, in his model, Woolfson introduces a force

$$f_r = \xi/r^m, \qquad (2.36)$$

where $\xi = GMr_0^{m-2}$, r_0 being the original value of r. Woolfson found that such a model star also loses mass from a bulge as predicted by the theoretical calculation.

However, in the computer simulation, he is able to follow the material after it has left the star. He finds that this material can move in orbits with a perihelion distance ranging from 31 Astronomical Units to 0·05 Astronomical Units, depending on when ejection from the distorted star occurs. He concludes that these limiting distances are in very good agreement with the limits of the planetary distances in the solar system. Woolfson also claims that much of the material that was disrupted could stay for a period close to the Sun, in this way acting as a resisting medium and so rounding off the orbits of the planets in such a way that the perihelion distances remain nearly constant.

Woolfson found that the mass of the planets that would form by direct contraction from the ejected material would be of the order of the mass of Jupiter. In his paper (Woolfson, 1964), he suggested that masses smaller than this critical value may contract, but in this case some evaporation would also occur. He does not elaborate very much on this point, and makes no real attempt to explain the chemical composition of the planets. It seems to the author that the mechanism suggested by McCrea & Williams (1965)—in connection with another theory, of which more details will be given in Chapter 4—whereby grains condense out and fall to the centre of these protoplanets, may be applied here. Near the Sun, it is only this core of grains that remains, giving both the correct composition and mass for the terrestrial planets. Further out, all the material remains, giving the more massive planets composed mainly of gas.

One of the objections to the tidal type of theory, namely that the Sun and the other star will both tidally disrupt the filament of material that has been ejected (Nolke's objection) was not considered by Woolfson. However, in a follow-up paper, Dormond & Woolfson (1971) do consider this point. Their conclusion is that, under certain circumstances, the tidal action of the Sun and of the other star virtually cancel each other out for a period sufficiently long to allow planetary condensations to form and evolve to a stage where they are capable of withstanding the tidal effects.

In this final form, and incorporating the mechanism of McCrea & Williams (1965), Woolfson's theory is capable of explaining most of the features of the solar system. Whereas the author, perhaps because of personal bias, prefers some other type of theory, he can see no fundamental fault in this theory. It certainly deserves to receive much more attention than it has received in the past from writers of reviews on the subject. Further comments on the virtues and faults of theories will be given in the concluding chapter; meanwhile, since Wolfson's theory is the latest theory in the family of tidal theories, we now turn our attention to other types of theories.

3. The Accretion Theories

3.1. Introduction

The theories which we now discuss are those that do not consider the formation of the Sun—it is simply assumed to have formed in whatever way stars form—and where the material from which the planets are to form is captured from interstellar space. That the Sun (or indeed any star) can capture material from interstellar space is fairly obvious, and detailed discussions of various aspects of such capture have been given by Hoyle & Lyttleton (1939), Bondi (1952) and Cremin (1969). What is given below is a somewhat simplified argument, which leads to conclusions almost identical to those given by the more detailed analysis, and which also gives an adequate estimate of the amount of material captured.

Consider a large, uniform, gas cloud of density ρ_0, in which the mean thermal velocity (average speed of the molecules—which is also the speed of sound in the gas) is W. Let a star of mass M move through this gas cloud with a constant velocity, V. There are two obvious simplifying cases that can be considered, $V \ll W$ and $W \ll V$. Take first the case when the speed of the star is much less than the mean thermal velocity. Then the mean speed of the gas molecules relative to the star can be taken to be W. If at any point W is less than the escape velocity from the star's gravitational field at that point, then the material there will be captured by the star. At a distance R away from the star, the escape velocity, V_E, is given by

$$V_E^2 = 2GM/R. \qquad (3.1)$$

Material can therefore be captured provided

$$W^2 < 2GM/R,$$

which means that all the material closer to the star than a distance R_c, where

$$R_c = 2GM/W^2, \qquad (3.2)$$

will be captured.

In general, there is a rate of flow of material of amount $\rho_0 W$ across unit area of any surface in one direction, which is balanced by a similar flow in the other direction. If, however, the gas on one side of such a surface is captured, then the flow becomes one-way. There is, therefore, a total rate of flow into the sphere of radius R_c of

$$4\pi R_c^2 \rho_0 W, \qquad (3.3)$$

and this material then also becomes captured. Substituting for R_c from equation (3.2), gives the accretion rate as

$$\frac{dM}{dt} = \frac{16\pi G^2 M^2 \rho_0}{W^3}. \qquad (3.4)$$

This expression agrees with that given by the more detailed calculations of Hoyle & Lyttleton (1939) for the same case.

For the situation where the star is moving very quickly compared with the mean thermal velocity, we may ignore the random motions of the gas molecules, so that material will now be captured if $V < V_E$. Substituting for V_E from equation (3.1) leads to a capture radius, R_c, for the case of

$$R_c = 2GM/V^2. \qquad (3.5)$$

Since the gas molecules are now considered to be at rest, the only new material that can enter the sphere of radius R_c is that which is swept up as this sphere moves through the gas cloud with a velocity, V, which is of course

$$\pi R_c^2 \rho_0 V, \qquad (3.6)$$

per unit time. Substituting for R_c from equation (3.5) gives the accretion rate for this case as

$$\frac{dM}{dt} = \frac{4\pi G^2 M^2 \rho_0}{V^3} \qquad (3.7)$$

which agrees well with the expressions found by Bondi (1952) and Cremin (1969).

The main dependence on velocity in their expression enters through the definition of R_c. Since this definition consists of a comparison of energies, this suggests that a possible expression, valid for all speed ranges, is given by

$$\frac{dM}{dt} = \frac{4\alpha\pi G^2 M^2 \rho_0}{(V^2 + W^2)^{3/2}} \qquad (3.8)$$

where α is a dimensionless constant with a numerical value of order unity.

As already mentioned, W is the mean thermal velocity of the gas, and is given by

$$W = 2\sqrt{\frac{2kT}{\pi m_1}}, \qquad (3.9)$$

where k is Boltzmann's constant, m_1 the mass of gas molecule (or atom, as the case may be) and T the temperature. For normal interstellar gas clouds composed mainly of hydrogen at a temperature of a few hundred degrees Kelvin, W has a value of about 10^3 m s^{-1}. The expected value for V, the relative velocity of the star and gas cloud, is also of the same order. If we take M to be the mass of the Sun, equation (3.8) gives the rate of accretion of mass as

$$\frac{dM}{dt} = 2 \cdot 5 \times 10^{32} \rho_0 \text{ kg s}^{-1}. \qquad (3.10)$$

If the mass of the matter to be accreted is planetary (say 10^{28} kg), and the accretion is occurring from a fairly dense interstellar cloud (say of density 10^{-18} kg m^{-3}), then equation (3.10) shows that the time required for such accretion is about 4×10^{13} s, that is $1 \cdot 2 \times 10^6$ years.

This is not an unduly long period on an astronomical time scale; so it appears to be perfectly feasible for a star like the Sun to accrete enough material to form the planets. One obvious objection to this very simple accretion picture is that the accreted material will fall radially on to the Sun. If planets are to be formed, it is therefore necessary to postulate some additional mechanism which is capable of generating, or transferring, angular momentum to the accreted material. It is at this point that the authors of accretion-type theories disagree with one another.

Lyttleton (1961), as well as generating arguments similar to the above, attempted to show that the initial rotation of the accreted cloud was sufficient to do this, but he did not produce any details of the planetary formation process. Since there is very little common ground between theories from now on, we shall discuss each one in turn.

3.2. Schmidt's Theory

We mention the following theory by the Russian astronomer, Schmidt (1944), not because it is still thought to be valid, but because it was perhaps the first to break away from the Jeans–Jeffreys type of cosmogony. Schmidt postulated that the Sun is not the only star that is passing through the gas cloud, but that there is at least one other star present. The function of this star is to introduce an asymmetry into the accretion problem, so that the accreted material has sufficient angular momentum to be capable eventually of forming the planetary system. Because of this angular momentum, the captured gas evolves into an elongated bun-shaped nebula rather than into a spherical configuration. The reason for this is obvious: namely, that rotational forces will assist pressure forces in overcoming gravity in the plane of rotation, but not perpendicular to this plane.

The Sun influences the gaseous nebula in other ways besides via its gravitational field, the most obvious being as a source of heat. It has already been mentioned that the equilibrium temperature at a distance r (in units of 10^{11} m) from the Sun is

$$T = 340/\sqrt{r},$$

indicating that the equilibrium temperature drops from about 400 K near the orbit of Mercury, to only 45 K in the outer regions of the solar system.

In the presence of the captured gas nebula, the temperature gradient will become much more severe than this, because the parts of the nebula nearest the Sun will absorb all the solar radiation, thus becoming hotter. At the same time, these parts shield the outer regions, which therefore become cooler. Calculating the exact temperature distribution under these conditions is very difficult, but there is no reason to believe that the crude picture suggested above is incorrect.

According to Schmidt, the effect of this enhanced temperature gradient is two-fold. Only the very non-volatile materials, essentially the silicates and the iron group, will be able to exist in a non-gaseous state in the high temperature region. These will still be found in grain form; and, paradoxically, the rate of growth of these grains may be quicker than under normal conditions, because the higher temperature implies that inter-grain collisions and grain–molecule collisions will be much more frequent. These grains are assumed to grow by cold-welding until they reach a size at which they are capable of gravitational capture of other grains. In this way, large objects may be formed. The solar radiation (the existence of the solar wind was not known when Schmidt proposed his theory) drives away the gas from the inner part of the system, thus leaving the existing type of iron–silicate terrestrial planet in this region.

Further away from the Sun, where the temperature is much cooler, grains again initially grow in much the same way, though there may now be grains of a slightly more volatile composition present as well. But when gravitational capture starts to play its part, the gaseous part of the nebula can now also be captured. This obviously leads to the formation of much larger objects, namely the major planets, eventually composed mainly of hydrogen and helium.

In the very outer regions of the system, though the temperature may be even lower, continual evaporation from the nebula will take place, especially of the lighter elements (hydrogen and helium). Consequently, when the planets eventually form there, they will be depleted in hydrogen and helium—the exact situation found in the solar system.

Much of the original work of Schmidt was qualitative, especially the discussion regarding the growth of planets. Though the argument does show that the general trend is to form a system similar in composition to the actual system, no indication is given of the likely number of planets, nor, therefore, of the mass of individual planets. Schmidt was the first in recent times to revive the old notion of Chamberlain (1901) and Moulton (1905) that growth of planets from small non-gaseous particles can occur. Since Schmidt's theory was proposed, other theories, to be described later, have also postulated a situation involving a solar nebula. Hence it seems more appropriate to discuss the modern advances in understanding the evolution of planets from such a nebula after discussing such theories. Although these newer ways of generating a solar nebula appear more appropriate, Schmidt deserves a mention for initiating the whole process.

3.3. Alfvén's Theory

One of Alfvén's main contributions to the solution of the problem of planetary cosmogony was to show that electromagnetic effects in plasmas could be important, both as a means of transferring angular momentum to the planetary material and as a mechanism for segregating the dust grains from the gas. Alfvén has carried out very extensive work on the subject of planetary formation, and space will only allow a very brief outline of it to be included here.

Alfvén was not, in fact, the first author to introduce electromagnetic effects into the problem. Birkeland (1912) suggested that ions, ejected from the Sun, would then spiral outwards towards their limiting circular orbits (the reverse of the process we shall describe below). Unfortunately, Birkeland never followed up this theory with detailed calculations. Berlage (1927; 1930a, b) had also used ejected ions in a theory of planetary formation. These ions were ejected into a flattened solar nebula which, in consequence, acquired a space charge: this was assumed to increase with distance from the Sun. On ignoring rotational forces (a very doubtful assumption), Berlage found that the equilibrium configuration was a series of concentric rings, each ring consisting of one particular element only, those of heaviest atomic weights being nearest the Sun. The planets then form from these rings, presumably by some nucleus forming and capturing the remainder of the ring.

Alfvén (1942a, b; 1943; 1946; 1954) considered the passage of the Sun through an interstellar gas cloud, with accretion of the cloud taking place in the manner which has already been described. But his main interest was in the actual motion of the captured material. If this material is at a distance r from the Sun, has a density ρ and an inward component of velocity v, then conservation of mass requires that

$$\frac{dM}{dt} = 4\pi r^2 \rho v. \qquad (3.11)$$

This equation shows that as r decreases, v can increase. If this is the case, then it seems reasonable to assume that collisions between atoms of the gas become more violent as the captured material proceeds towards the Sun. In a conventional sense, this would correspond to an increase in the temperature of the gas. The obvious consequence is that the atoms become ionized, and it therefore becomes necessary to consider the motion of charged particles through a magnetic field. In reality, this is a plasma problem, but a very reasonable idea of the outcome can be obtained by considering a single particle of mass m and charge q moving under the gravitational and magnetic effect of the Sun. For simplicity, let us assume that the solar magnetic field is generated by a dipole of moment $\boldsymbol{\mu}$ and that the particle is initially moving in the plane defined by this vector $\boldsymbol{\mu}$. The problem then becomes two-dimensional, since all the forces are also acting in this plane. Using polar coordinates (r, θ), the equation of motion of this particle becomes

$$m\ddot{\mathbf{r}} = -\frac{GM_\odot m}{r^2}\hat{\mathbf{r}} + \frac{q}{r^3}(\dot{r}\hat{\mathbf{r}} + r\dot{\theta}\hat{\boldsymbol{\theta}}) \times \boldsymbol{\mu} \qquad (3.12)$$

since the magnetic field in the plane is given by $\boldsymbol{\mu}/r^3$. Since $\boldsymbol{\mu}$ is perpendicular to both $\hat{\mathbf{r}}$ and $\hat{\boldsymbol{\theta}}$, equation (3.12), written in component form, gives

$$m\ddot{r} = -\frac{GM_\odot m}{r^2} + \frac{q\mu\dot{\theta}}{r^2} + mr\dot{\theta}^2 \qquad (3.13)$$

and

$$\frac{m}{r}\frac{d}{dt}(r^2\dot{\theta}) = -\frac{q\mu\dot{r}}{r^3}. \qquad (3.14)$$

Equation (3.14) can be integrated to give

$$mr^2\dot{\theta} = -\int\frac{q\mu\,dr}{r^2} = \frac{q\mu}{r} \qquad (3.15)$$

on inserting a boundary condition that at large distances from the Sun, there is no angular momentum in the particle relative to the Sun.

Substituting from equation (3.15) into equation (3.13) for $\dot{\theta}$ gives

$$m\ddot{r} = -\frac{GM_\odot m}{r^2} + \frac{2q^2\mu^2}{mr^5},$$

or $\qquad (3.16)$

$$\dot{r}\frac{d\dot{r}}{dr} = -\frac{GM_\odot}{r^2} + \frac{2q^2\mu^2}{m^2 r^5}.$$

This equation can also be integrated to give

$$\dot{r}^2 = \frac{2GM_\odot}{r} - \frac{q^2\mu^2}{m^2 r^5}, \qquad (3.17)$$

on inserting the boundary condition that $\dot{r} = 0$ at large r.

The interesting feature of this equation is that there exists one other value for r at which $\dot{r}$ is zero, namely

$$r_c = \left(\frac{q^2\mu^2}{2GM_\odot m^2}\right)^{1/3}. \qquad (3.18)$$

The particle can therefore never approach the Sun closer than this distance r_c.

It is obvious that the value of r_c depends only on the ratio q/m, that is the ratio of charge to mass for the particle. There are two types of particle that may reasonably be expected to be accreted by the Sun—a typical gas atom (generally hydrogen) and a dust grain. When a hydrogen atom becomes ionized, its charge-to-mass ratio is well defined, and inserting this value together with an estimate for the solar magnetic field yields a value for r_c of about 10^{13} m. The real value should be somewhat smaller than this since ionization does not, in fact, take place until the speed of the particle is about 5×10^4 m s^{-1}, so that the boundary condition in equation (3.17) becomes somewhat modified. Alfvén concluded that when account is taken of this, the gas will reach minimum distance in the region now occupied by the major planets.

For the dust grains, the charge-to-mass ratio will be much smaller. Roughly, only the surface layers of these will become ionized, and so one would expect the charge-to-mass ratio to be proportional to the ratio of surface area to volume, that is inversely proportional to the radius. Now the radius of a grain is about 500 times that of a proton, and as equation (3.18) shows that r_c is dependent on $(q/m)^{2/3}$, the distance of closest approach will be reduced by a factor of somewhat less than a hundred. This means that the grains will reach their minimum distance in the region roughly corresponding to that now occupied by the terrestrial planets.

Alfvén thus postulates that the planets formed not from one nebula but from two. One of these nebulae forms in the region of the terrestrial planets, and is composed mainly of the non-volatile grains (the correct composition for the formation of the terrestrial planets). The other nebula occupies the region now containing the major planets, and is composed mainly of hydrogen (again the correct composition for a major planet). Alfvén, in fact, claims that the density distribution in the existing solar system, with the relative lack of material in the region of the asteroid belt, suggests strongly that the planetary system could not have developed from one nebula. Alfvén's mechanism for the process of accretion provides the material from which the planets form with the correct chemical composition for the main parts of the system.

Together with Arrhenius (1970a, b; 1973), Alfvén has also considered the actual growth of planets from the nebulae. They have investigated the aggregation into larger bodies of a swarm of small, solid grains initially moving in Keplerian orbits. Alfvén (1969; 1970; 1971) claims that collisions between such particles do not lead to a spreading of the particles but rather to an equalization of the orbits, provided that the collisional frequency is smaller than the orbital frequency (less than one collision per orbit). Such a stage will inevitably occur, if not initially, then after accretion has reduced the number of particles, so that collisions are more infrequent. The outcome of such an occurrence is to focus the particles into streams—jet streams as they have been called by Alfvén—that is, into orbits with increasingly similar orbital elements and velocities. Eventually in such a scheme, some embryos will be formed which will grow to such a size that they are capable of gravitationally capturing the remaining particles. It is claimed that the distribution of matter in the asteroidal belt, and the existence of meteor streams, gives support to this jet-stream hypothesis.

Though Alfvén's theories do not give an exact quantitative analysis of the final stages of the accumulation into planets, and thus do not provide an exact prediction of the individual masses and orbital positions of the planets, it is fair to say that some explanation is offered for most of the major features of the solar system.

3.4. The Theory of Pendred and Williams

Pendred & Williams (1968) consider that the mass of gas and dust from which the planets are to form is captured as the tail end of the process of the formation of the Sun. In a very strict sense, this theory does not belong to the category being discussed in this chapter. However, the Sun plays no essential part in the process, and since, in general, this theory displays many of the characteristics of an accretion theory, it was thought advantageous to discuss it here.

The Sun is thought to have formed (in whatever way stars form) from a gas cloud in the Galaxy. After this event, there will still be a considerable amount of gas around, relative to which the Sun will have negligible velocity, but which will have some rotation. (All objects in the Galaxy have some rotation due to galactic rotation of magnitude about 10^{-15} s^{-1}.) This gaseous material will be accreted by the young Sun; but the accretion process is now different from any that have been considered so far, because of the initial rotation of the cloud. We assume that angular momentum is conserved in the system, which means, in this case, that angular momentum is conserved both about the axis of rotation of the cloud through the Sun, and also about the Sun itself. The gas is assumed to behave adiabatically during its motion. An exact analysis of this motion of the accreted gas is very difficult, and was not attempted by Pendred & Williams, but considerable insight into what happens can be obtained in the following way.

The gas nearest to the Sun will fall into it, and so, no doubt, contribute to its rotation. However, as gas from greater distances is accreted, it will possess higher angular momentum, and will therefore not fall into the Sun but instead reach a perihelion at some point external to it. It is this material that is clearly of interest from the point of view of planetary cosmogony. Consider a ring of this material, originally at some distance R_0 from the axis of rotation (denoted by AB in Figure 3.1). In isolation, each particle of this ring would pursue an elliptical orbit about the Sun under the influence of the solar gravitational field, whilst at the same time conserving angular momentum about the axis of rotation. In isolation, therefore, the ring as a whole would contract to some other ring CD on the opposite side of the equatorial plane (defined by the rotation axis) before returning to AB. It is, however, clear that the gas as a whole cannot behave like this, since there is a ring identical to AB situated initially below the equatorial plane which would wish to contract to a ring C'D' identical to CD but situated above the equatorial plane. Both these rings would reach the equatorial

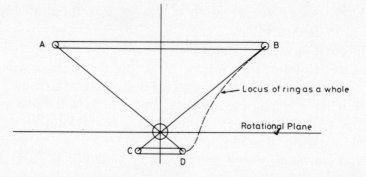

Figure 3.1. The formation of a high density ring according to Pendred and Williams.

plane at the same place and time, and it is clearly absurd to assume that they could pass through one another without interacting. The obvious outcome is that the paths of both rings are modified such that they reach perihelion in the equatorial plane. Because of the need to conserve angular momentum about the axis of rotation, the modified perihelion distance must be about the same as the distance of the undisturbed perihelion from the axis of rotation. The outcome, therefore, is to produce temporarily a ring of high density in the equatorial plane. As time progresses, material from greater distances will reach the equatorial plane and form a high density ring there. An external observer will therefore see a ring of relatively high density forming in the equatorial plane close to the Sun, which will gradually appear to move out to greater distances (though in reality it is a succession of new rings forming).

The temperature of a ring is governed by two effects, the heating of the Sun and the internal energy of the adiabatic gas. Both these effects indicate that the temperature will be highest in the rings closest to the Sun. Pendred & Williams argue that condensation can easily occur in this high density ring, and so the most non-volatile (refractory) materials condense out first. Though hydrogen and helium will never be able actually to condense, it is arguable that the internal energy is low enough in the major planet region to allow considerable amounts to be captured once sizeable condensations have formed.

In an attempt to be more quantitative than the above qualitative discussion, Pendred & Williams introduced an arbitrary expression for the rate of flow of material into the system as

$$F = \frac{H}{p_0^{1\cdot 1}} \qquad (3.19)$$

where p_0 is the perihelion distance of the material, and H is a constant. When reasonable numerical values are introduced for all the parameters, Pendred & Williams found that the high-density ring would reach the orbit of the Earth in about 10^5 years, and that the apparent outward speed of the ring is then about $0\cdot 2$ m s^{-1}. It also transpires that the mass of material condensing in this region is about 10^{25} kg, assuming that only the non-volatiles condense. In the major planet region, the volatiles are also captured, and the mass will be increased by a factor of order 100. The agreement with the actual system is fairly reasonable in these numerical values.

In a second publication, Pendred & Williams (1969) attempted to determine the mass and orbital radius of the planets. They introduced the concept of gravitational reach, that is the distance to which a condensation is able to capture other condensations. They then assumed that the first condensation occurred at an arbitrary distance r_0, the second condensation would then start at a distance $r_0 + L$ where L is the reach of the first and so on. In this way, Pendred & Williams obtained the planetary masses and orbital distances given in Table 3.1. Also shown are

Table 3.1

Planet	Observed Mass	Calculated Mass	Observed Orbital Radius	Calculated Orbital Radius
Venus	0·82	3·2	0·72	0·56
Earth	1·00	1·01	1·00	1·00
Mars	0·11	0·45	1·5	1·40
Jupiter	318	318	5·2	5·1
Saturn	95	83	9·5	9·8
Uranus	15	38	19·0	14·0

(All quantities are expressed in terms of Astronomical Units)

the actual values for the planets for comparison. We see that there exists quite reasonable agreement.

Pendred & Williams assumed that all the accretable material in any ring was, in fact, accreted, and they said nothing about the actual capture mechanism (and gave no details of the growth process). The actual agreement between observations and prediction in this theory is impressive; but it has to be remembered that while the qualitative argument appears sound, in order to quantify it many assumptions were made and the free parameters were suitably selected. Too much weight should not therefore be given to the numerical agreement obtained, though the lack of detail in the final growth of a planet is no greater than in many other theories.

4. Processes for the Formation of the Sun and Planets

4.1. Introduction

The final category in our classification of theories covers those in which the formation of the planets takes place as a direct consequence of the formation of the Sun. This type of theory therefore attempts to answer a somewhat wider question than planetary formation: that of the formation and early evolution of stars. Though such theories do not predict a planetary system surrounding every star, planetary systems would be a wide-spread phenomenon if they were correct, certainly being far more common objects than if either of the other two types of theories proved to be correct.

We shall not discuss here all the known facts concerning star formation or early stellar evolution (a brief account of the latter has been given in Chapter 1), but will only mention such points as are relevant to the planetary formation process. It is obvious that, as we know of the existence of many more stars than planets, in theory it should be easier to observe the formation of stars occurring and, hence, determine the correct method of star formation. Though observations have been made of very young stars, it is fair to say that at present all the theories which we shall mention are consistent with the known facts regarding star formation.

It is, perhaps, a surprising fact that the early theories of Descartes (1644), Kant (1755) and Laplace (1796) could all be placed in this category, but that it is only since the end of the Second World War that this type of theory has become respectable. Today these theories are the ones most commonly accepted, and, in consequence, considerable recent work has been carried out on them, though some of this work has been concerned only with specific aspects of a theory. It seems, therefore, more appropriate to discuss ideas, rather than to give an account of X's theory followed by Y's theory. Broadly speaking, the theories divide naturally into three types: those following the ideas of Descartes, Kant and Laplace, respectively, and we shall discuss them in this order.

4.2. Descartes' Theory and its Successors

In 1644 Descartes postulated that the Universe (remember that this was before galaxies and stellar clusters had been discovered) was filled by ether and matter. Then as soon as any motion was imparted, vortices were set up, so that the Universe became filled with circular eddies of all sizes. The friction between the eddies smoothed down the rough shape of the primordial matter and any matter "filed" away in this process was assumed to tend towards the centre of the vortex, thus forming the Sun, while the coarser bodies were captured in the vortex and formed the planets. Around these planets, secondary vortices formed, in which the satellites were captured.

Judged by modern-day standards, Descartes' theory is not much of a theory, but it has to be remembered that at the time it was formulated, Newton had not proposed his law of gravitation. It was also only a few years since Galileo had had his troubles with the Church authorities, and it was somewhat safer not to make one's support for Copernican ideas too obvious.

Turbulence plays a part in many modern cosmogonical theories, which, in this respect, therefore follow Descartes. The first of these theories was by von Weizsäcker (1944) who assumed that the Sun had an extended envelope surrounding it after formation. Because of its rotation, this envelope will flatten in much the same way as was described by Schmidt (1944) and others. The mass of the envelope is taken to be about a tenth of a solar mass, and the mean density is 10^{-7} kg m^{-3}. The Reynolds number in the disk is, therefore, very large (of order 10^{11}), and the material will be turbulent. Von Weizsäcker argues that pressure and frictional forces are small, so that each mass element will follow a Keplerian orbit. If there exists a family of such orbits, all with the same perihelion, but with differing small eccentricities, then relative to an element in a circular orbit ($e = 0$) these other elements will appear to be moving in small ellipses. For larger perihelion distances, the rotation period around the secondary ellipses increases. Von Weizsäcker pictures this situation as a vortex circling the Sun.

Of course, many such vortices will be formed, and it is envisaged that a quasi-stable arrangement of vortices will form into a series of rings, each of them consisting of a number of vortices touching each other. For the good reason that it gives the best answer, von Weizsäcker postulated that there were five vortices per ring: this gives a tolerable approximation to the Titius–Bode law for planetary distances.

All these vortices will be rotating in the opposite direction to the general motion about the Sun; so there is considerable friction at the points of contact between vortices, necessitating the presence of secondary 'roller-bearing' vortices. The total mass of these roller-bearing eddies is about one-fifth of the total mass of the disk, leaving four-fifths in the main vortices.

Von Weizsäcker assumed that in any gas there would exist nuclei for condensation which would grow in size by the capture of smaller nuclei. As soon as a certain size is reached, this straightforward capture gives way to gravitational capture. Gravitational growth proceeds much more rapidly than the straightforward growth: it is estimated that objects of planetary dimensions can be formed in this way in a period of about 10^8 years.

It is assumed that all the particles share in the turbulent motion with the exception of the very large ones. In this case, the condensation will take place mainly in those regions where the mean free path of the turbulence is small, since the number of collisions will be highest there (i.e. in the roller bearing eddies). Since the 'roller bearings' occur in rings separating the main vortices, and since these vortices increase in size with increased distance from the Sun, condensation therefore occurs in rings whose distances from the Sun approximately agree with the Titius–Bode law. Von Weizsäcker assumed, but did not prove, that eventually only one condensation will survive in any ring, forming a planet.

There will, of course, be some exchange of angular momentum and energy between rings. As a consequence, von Weizsäcker suggested that the heaviest elements would fall towards the Sun and the lightest would be driven further away, thus offering some explanation for the composition of the planets.

One obvious criticism of this theory is that undisturbed Keplerian orbits are impossible: one has to consider the problem hydrodynamically, and a realistic turbulence theory suggests that dissipation of the disk will occur on a much shorter time scale than 10^8 years, so that there will be no time for condensations to grow.

Ter Haar (1948; 1950) criticized a number of points in von Weizsäcker's theory and modified it accordingly. The regular eddies assumed by von Weizsäcker were discarded, and replaced by random turbulence. This leads to a very thick nebula in which gravitational instabilities cannot easily occur. Consequently, the planets must grow by collisional accretion.

Ter Haar made use of the heat output from the Sun, pointing out that the inner parts of the nebula would be considerably hotter than the outer parts, so that only the non-volatiles would condense close to the Sun, while almost all compounds could do so in the outer regions. Much larger bodies would therefore form in the outer regions, and ter Haar suggested that they become capable of gravitational accretion before turbulence dissipates the nebula. However, the time required for turbulent dissipation of the nebula is only 1000 years, and it is difficult to see how any appreciable growth can occur in such a short interval of time.

It is possible that magnetic fields could stabilize the nebula against dissipation by turbulence: Prentice & ter Haar (1969) have shown that cosmic rays could maintain an ionization level in the nebula high enough for magnetic fields to be important. Unfortunately, a detailed study of the effects of magnetic fields has not been carried out, so it is not clear whether they hinder or assist dissipation.

Kuiper (1949; 1951a, b) was also attracted by von Weizsäcker's ideas concerning a turbulent nebula but, like ter Haar, rejected the regular vortices postulated by von Weiszäcker. He thought that large gravitational instabilities could form in the nebula.

We have already shown in Chapter 2 that a spherical mass M, with a density ρ and radius R, at a distance $r \gg R$ from the Sun can hold together only if

$$\rho \sim \frac{M}{R^3} > \frac{M_\odot}{r^3}. \tag{4.1}$$

Kuiper showed that for large areas of the nebula, the density would exceed the value given by equation (4.1), so he concluded that blobs, able to hold together, would form. However, if the density is so high, then account must be taken of the gravitational field of the nebula in order to arrive at a criterion like that given in equation (4.1), and this has not been done. Also, equation (4.1) is concerned only with the blob density and says nothing about the size or mass of this object, yet Kuiper assumed that protoplanets would be formed, one corresponding to each of the present planets. These protoplanets would all be of solar composition, and so segregation of the condensables (which we shall discuss shortly) is necessary in order to form the terrestrial planets.

Whipple (1946; 1948) also made use of protoplanets, within which the segregation of the non-volatiles to form the terrestrial planets occurred. In Whipple's picture, the Sun formed by the contraction of a gas cloud possessing very little angular momentum, which subsequently captured a second smaller cloud. In this second cloud, partially condensed sub-clouds are assumed to exist. These sub-clouds become the protoplanets, which eventually move in nearly circular orbits due to the influence of a resisting medium. Unfortunately, most of the features of the solar system are introduced as *a priori* postulates in Whipple's theory.

Edgeworth (1946) demonstrated that the planets could not form by direct condensation from a gas cloud. This was much the same conclusion as had been reached much earlier by Jeffreys (1918), but he had considered more general density distributions. In a second paper, Edgeworth (1949) therefore had to propose that the formation process did not involve such direct condensation. He postulated instead that, during the formation of the Sun, some material is captured in orbit about the Sun. Because of collisions between different parts of this material, it will eventually form a disk about the Sun whose plane is coincident with the plane of rotation. Edgeworth recognised that the solar rotation had to be reduced but did not know of the solar wind as it had yet to be discovered. He decided that magnetic fields were too speculative, and so there remained only viscous forces. These viscous forces cause eddies to form in the disk, while collisions between them will increase their mean size until eventually they have planetary dimensions. Edgeworth estimated that the time required for this to occur is of the order of 10^5 years for objects near the Earth's orbit, but increases to about 5×10^6 years near Neptune's orbit. He suggested that the planets formed in two sequences, one containing the planets from Mercury to the asteroids, and the other containing the planets from Jupiter to Pluto. He envisaged the asteroids as being the tail-end of one sequence, and the comets as the tail-end of the other. Whereas this may be true in the case of the asteroids, the necessary mean distance for comets is so large as to make it unlikely that they can be the tail-end of any such sequence. The first sequence must form from refractory material and the second from predominantly hydrogen material. The chemical differences therefore arise almost as a postulate rather than as a consequence of any calculation, though doubtless an argument concerning solar heating could be used. The main weakness of the theory is that viscous friction can only be effective in slowing down the Sun if the material in the disk actually touches the Sun at all times. Hence one would expect to find material much closer to the Sun than Mercury's orbit. It is, indeed, very doubtful if viscous friction could be very efficient as a slowing-down mechanism for the Sun, even if material did touch the Sun.

Some of the ideas of Edgeworth, however, reappear in the most recent of the theories that broadly follow Descartes. This theory was proposed by McCrea (1960a), and also incorporates both the ideas of turbulence and protoplanets. The first paper was mostly descriptive: McCrea (1960b) published further details and calculations shortly afterwards. He first considered the problem of star formation and, since most stars appear to have belonged to a stellar cluster, postulated that stars form in clusters. Accordingly, he envisaged an interstellar gas cloud with a mass of about 100 solar masses as the initial state. This cloud is assumed to have contracted to a density of about 4×10^{-9} kg m^{-3} by a process described by him earlier (McCrea, 1957). The gas cloud is assumed to be composed mainly of molecular hydrogen (a rather startling assumption at the time it was made, but now fully vindicated by recent observations of the interstellar medium) which means that it is a very efficient radiator of energy. In consequence, the temperature remains at about 50 K throughout. It is at this point that McCrea departs from the usual treatment of the contraction of a gas cloud, for he maintains that such a cloud would be supersonically turbulent. Very little is known about such a phenomena, so McCrea postulated that the following model would represent it reasonably well (no doubt inspired by the mixing-length theory for representing convection). The cloud is taken to be broken up into a chaotic swarm of cloudlets, or *floccules* as McCrea termed them. These floccules are continually dispersing, with new ones reforming, and so they cannot be gravitationally-bound objects. This means that their gravitational energy must be less in modulus than the thermal energy, that is

$$\frac{3}{5}\frac{GM_f^2}{R_f} < \frac{3}{2}\mathscr{R}TM_f, \qquad (4.2)$$

where M_f is the mass of a floccule, R_f its radius and $\mathscr{R}$ the gas constant. Replacing the radius by the mean density, ρ_f, gives

$$M_f < \left(\frac{5}{2}\frac{\mathscr{R}T}{G}\right)^{3/2}\left(\frac{4}{3}\pi\rho_f\right)^{-1/2}. \qquad (4.3)$$

McCrea takes the random speed of the floccules to be the speed of sound, namely 10^3 m s^{-1}. He also assumes that the mean free path of the floccules, L, is 5×10^{12} m, and that about 10^5 floccules exist within a sphere of radius L. He admits that this list of assumptions is rather long, but points out that all other features of the model are now determined.

The mass within the volume of radius L is, of course,

$$\tfrac{4}{3}\pi L^3 \bar{\rho} = 2 \times 10^{30} \text{ kg} \tag{4.4}$$

on using the given numerical values. Since there are 10^5 floccules in this volume, it follows that the mass of a floccule is $M_f = 2 \times 10^{25}$ kg, while the definition of the mean free path gives the radius of a floccule as just under 10^{10} m. The density of a floccule, ρ_f, can then be calculated as $6{\cdot}7 \times 10^{-6}$ kg m^{-3}.

These floccules will collide amongst themselves. In the collisions, the parts directly concerned will coalesce, while the other parts are lost to the floccule. Collision can therefore either produce a larger than normal, or a smaller than normal, floccule. If there is a succession of successful collisions, then a much larger floccule will form. Equation (4.3), when the value for ρ_f is inserted, shows that an 18-floccule mass can hold together gravitationally. From time to time it seems likely, therefore, that such a stable condensation will form, and become a permanent feature in the cloud. McCrea argues on purely statistical grounds that one, and only one, such condensation will form within a sphere of radius equal to one mean free path. Further, in time, these condensations will capture most of the available mass. The average mass of the object formed by this process is the mass within the sphere of radius L given by equation (4.4) as 2×10^{30} kg, namely a solar mass. McCrea concludes that stars are formed in this way. It should be noticed that any growing condensation will tend to capture those floccules moving directly towards it, and so the stars formed by this process will tend to have low angular momentum.

Williams (1969) attempted to make the numerical values for the floccules adopted by McCrea less arbitrary. He made a few postulates regarding the behaviour of floccules, and showed that a cloud of mass 100 solar masses would indeed fragment into floccules having just the properties outlined by McCrea. An interesting consequence of Williams' work was that it showed that if a massive cloud with a mass similar to a galactic nucleus contracted, then very massive stars, such as those possibly required to explain quasars, would form.

Williams & Cremin (1968) investigated the evolution of a young star that was acquiring mass in the way envisaged by McCrea: an initially small mass increasing as floccules (of finite mass) were added to it. They showed that such evolution was consistent with the known facts regarding the evolution of young stars. McCrea's theory for star formation thus seems plausible, but it is the formation of planets that is of interest to us.

Not all the floccules will be accumulated by any protostar: those whose undeflected paths would pass relatively far from any star will be captured in orbit about a star. McCrea estimates the number of these captured floccules by considering the angular momentum. The random walk theory tells us that the sum of N random vectors is approximately $(N)^{1/2}$ times the magnitude of one vector. The initial angular momentum contained in a sphere of radius L is thus approximately

$$M_f(N)^{1/2}VL = 5 \times 10^{13} M_f \, \text{m}^2 \, \text{s}^{-1}.$$

The final configuration consists of N' floccules moving in orbit at some mean distance, R. By equating the two amounts of angular momenta, McCrea estimates that about 1000 floccules will be captured in orbit. These floccules will collide and accumulate in the same way as did those which formed the stars. Now, however, a new effect has to be considered. If, as a result of a collision, a new floccule is formed which is low in angular momentum, this composite floccule will fall into the Sun. In this way, retrograde floccules are eliminated. At the same time, the collisions will tend to flatten the system into one plane, and McCrea claims that the final result is a near-coplanar system composed of about 200 floccule masses. Since approximately 20 floccules form a stable core, he suggests that the final captured mass is in the form of about 10 protoplanets, each about 20 floccules in mass. These protoplanets therefore have the general characteristics of major planets.

The problem of the settling of the floccules and the elimination of retrograde orbits has been the topic of a number of investigations. In a review article, Woolfson (1969) mentions that with a probability of 0·6 of having a prograde orbit, as deduced from McCrea's work (from an initial 1000 floccules, 400 retrograde ones destroy 400 prograde ones, leaving behind 200 prograde), there is a very real chance of producing a planet moving in a retrograde orbit.

Williams & Galley (1971) investigated this contention in more detail using a stochastic model. The problem they investigated is identical to the following. Take a bag containing 1000 balls. There is a probability 0·6 that any ball is red and 0·4 that it is black. Balls are taken out one at a time and placed in a pile. If there are 20 balls in a pile, a new pile is started. If at any stage the number of red balls in a pile is equal to the number of black balls, the pile is discarded (low angular momentum falling into the Sun). The question is how many red piles can be formed, and how many black. The answer they obtained was that the probability of forming a black pile was small, but that a number of red piles could be formed. Hence it appears unlikely that a retrograde orbit for a planet will exist. Williams & Galley also showed that the expected orbital distances of the planets roughly corresponded to those of the actual planets, the distances being obtained from the red-to-black ratio which gives the angular momentum.

Aust & Woolfson (1971) investigated both the settling and the elimination of retrograde orbits using a three-dimensional computer model. They generated 1000 random vectors, corresponding to the angular momenta of each of the 1000 floccules. These were added together, attempting to form a 20-floccule object, while testing to see whether the angular momentum of the composite body was small enough to allow it to fall into the Sun. They concluded that many retrograde planets would be formed, and that the orbits would not be anywhere near coplanar.

Williams (1972) pointed out some of the shortcomings of the model of Aust & Woolfson, but it was left to Williams & Donnison (1973) to investigate the problem with an improved model. The main improvement stemmed from the recognition that the collision rate between floccules moving in opposite directions is much higher than for those moving in a similar direction. They also changed the addition technique so that a multi-floccule agglomeration would collide with another such agglomeration, rather than with one floccule as in the Aust–Woolfson model. They concluded that there was a real tendency both for flattening and for the elimination of retrograde orbits, but that a large number of collisions was necessary to generate the degree of flattening found in the present solar system. Since all the collisions are not successful, such a large number of collisions may have occurred in the scheme envisaged by McCrea.

At this stage all the protoplanets are similar to each other in composition and mass. In order to produce the terrestrial planets, it is therefore necessary to segregate the terrestrial elements from the hydrogen and helium and remove the latter. The obvious segregation process is gravitational settling, the heavier grains falling towards the centre of the protoplanet. This process was investigated in some detail by McCrea & Williams (1965), who concluded that normal interstellar grains could not segregate in the time available, but that the time of fall became reasonable if the falling grains were assumed to accrete all other grains with which they came into contact. At the time this 'sticky' property of the grains was pure hypothesis but experimental work by many authors (e.g. Kerridge & Vedder, 1972) now shows that this is likely to occur. We give below a simplified form of the mathematics for the case most likely to be of interest.

If the grain is growing, it soon reaches a size where the usual Stoke's formula gives the correct resistance to motion, namely

$$F_{\text{res}} = 6\pi\eta r\dot{x} \tag{4.5}$$

where η is the coefficient of kinematic viscosity, r the grain radius and x the distance fallen. In a spherical protoplanet of density ρ and radius R, the equation of motion therefore becomes

$$\frac{d}{dt}(m\dot{x}) = \tfrac{4}{3}\pi G\rho(R - x)m - 6\pi\eta r\dot{x}, \tag{4.6}$$

where $m = (4\pi r^3 \rho_g)/3$ is the mass of a grain.

Baines, Williams & Asebiomo (1965) have found an expression for the radius of a grain which accretes all other grains: for the case of interest, this gives

$$r = r_0 + \frac{\lambda \rho x}{4\rho_g} = r_0 + \alpha x \text{(say)}, \qquad (4.7)$$

where r_0 is its initial radius and λ the proportion by weight of grains in the gas. Since any falling object tends to fall at near its terminal velocity, an approximate solution to equation (4.6) is given by integrating:

$$6\pi\eta r \dot{x} = \tfrac{4}{3}\pi G \rho (R - x) m. \qquad (4.8)$$

This equation can be simplified to give

$$\dot{x} = K(r_0 + \alpha x)^2 (R - x) \qquad (4.9)$$

where

$$K = \frac{8\pi G \rho \rho_g}{27\eta}.$$

Integrating equation (4.9) from $x = 0$ to $x \simeq R$, and keeping the dominant term only, gives the time of fall of a grain as

$$\tau = \frac{27\eta}{2\pi G \rho^2} \lambda R r_0. \qquad (4.10)$$

Substitution of the appropriate numerical values shows that the segregation time is of the order of a few hundred years. Williams & Crampin (1971) reinvestigated the same problem, but used numerical techniques for solving the correct equation of motion rather than approximating in order to obtain an analytical solution. They confirmed that the segregation time is indeed short. Both the above investigations assumed for simplicity that the protoplanet was of uniform density throughout. Williams & Handbury (1974) dispensed with this assumption and considered centrally-condensed protoplanets. They showed that for this situation the segregation time is again short. Therefore it seems reasonable to conclude that a core, essentially composed of elements other than hydrogen and helium, can form in the interior of a protoplanet of the type envisaged by McCrea (or Woolfson (1964), from his tidal theory).

In order to produce a terrestrial planet, we must do more than this: a mechanism for the removal of the outer layers of hydrogen and helium must also be present. There appear to be two obvious mechanisms for doing this. Firstly, we have already mentioned that any object placed in the gravitational field of the Sun experiences a disruptive tidal force; it is only if the gravitational field of the object itself is greater than the tidal force that the object can hold together. The exact formula giving the relationship between quantities of interest depends on the details of the model, but they all have the same form, namely

$$\frac{M_\odot}{R^3} < \frac{\alpha M}{r^3} \qquad (4.11)$$

where R is the distance from the Sun, M and r the mass and radius of the object, and α a dimensionless constant whose numerical value is between 1 and 3. The equation can be rewritten in the form

$$R > R_\odot \left(\frac{\alpha \rho}{\rho_\odot}\right)^{-1/3}. \qquad (4.12)$$

Equation (4.1), is obviously one of this family of equations. It can be interpreted as saying that a protoplanet of a given density can hold together only if its distance from the Sun is greater than a critical value given by equation (4.12). With the assumed protoplanet density of 6.7×10^{-6} kg m^{-3}, this critical distance turns out to be comparable with the orbital distance of the asteroidal belt. Any protoplanet orbiting at a greater distance than this should therefore contract as a whole to form a major planet, while any orbiting interior to this distance should be disrupted, leaving only the core, since this is at a higher density.

Of course, it cannot be that simple, since the disruptive process is very quick; if the protoplanets that are to form the terrestrial planets had always orbited interior to the asteroidal belt, they would have been disrupted before any core could form. Presumably the protoplanets formed in highly eccentric orbits, but these were rounded off by the large amount of free gas initially present. At first they would spend most of their time external to the critical distance, and a core could form at this epoch. It is only later, after rounding has occurred, that most of the disruption takes place.

The second possibility is that the solar wind is responsible for removal of the envelope in the case of the terrestrial planets. It injects both energy and momentum into the gas, which can lead to the evaporation of more gas from the protoplanet than was carried into it by the wind. Williams & Handbury (1974) have investigated this phenomenon, and have concluded that if the wind was enhanced in the early Sun to much the same level as in T-Tauri stars (which are observed to lose considerable mass at present), then the envelope could be removed for all planets out to, but not including, Jupiter.

McCrea's theory can thus account for the chemical composition of the planets, as well as their mass and general characteristics. One other relevant investigation is by Donnison & Williams (1974). They considered the contraction of the protoplanet that was to become Jupiter, using the normal equations of stellar evolution. They concluded that there is nothing in Jupiter's present state which contradicts the existence of such a contraction stage.

Finally, it should be mentioned that Urey (1966) considered that many of the points given above must, indeed, have occurred from the cosmochemical evidence. In order to form diamonds such as those found in some meteorites, Urey maintained that objects of at least lunar size must have formed (Urey means lunar-sized after the removal of gas). Collisions occurred between these objects, destroying most of them, but in the process aligning them in a plane. Close to the Sun, only the non-volatiles accumulated from the debris of these collisions to form the terrestrial planets, while further away gas was also captured.

Thus, starting from the early theory of Descartes, we see that modern theories are capable of explaining most of the observed features of the solar system.

4.3. The Theory of Kant and its Successors

Kant (1755) assumed that the universe was initially filled with gas, distributed roughly uniformly, but with slight density variations occurring in it. He postulated that these higher density regions would act as nuclei for condensation, and so grow, rather than be smoothed out, with the passage of time. The condensations would then gravitationally contract, eventually forming stars. In the initial state, the condensations were assumed to be rotating, so that conservation of angular momentum must mean that rotational forces become important in the development of the condensation. Kant therefore maintained that all the condensations became disks flattened in the plane of rotation. Secondary condensations formed in each such disk, and the planets developed from these, whilst the primary condensation in the central regions formed the Sun.

Kant assumed that the lines of development outlined above necessarily follow from one another; he did not, for example, enquire under what conditions secondary condensations could form in the disk. Nor did he obtain numerical values for the mass and orbital distances of the planets and for their chemical composition. Nevertheless, for such an early theory, it is very impressive. Not surprisingly, many modern theorists have taken it as their model, and investigated the development of such a collapsing rotating cloud in much more detail.

There appears to be no doubt about the assertion that stars form from interstellar clouds, and that, in general, many stars form from the same cloud. One can point to regions, particularly in the Orion nebula, where star formation seems to be taking place at present. But there is some problem about the fragmentation of an interstellar cloud into stellar masses: one suggested solution is that proposed by McCrea and already described. Nevertheless, it is presumably permissible to postulate that such a fragmentation can occur, and to attempt to follow the evolution of such a gas cloud (of mass approximately one solar mass) as it contracts from low densities. Discussion of the fragmentation process belongs to the study of star formation, and is therefore outside the strict field of interest of this account.

Let us consider a gas cloud of mass M and initial density ρ_0. For simplicity, we take the case when the cloud always remains spherical with an initial radius, r_0. Let us also ignore all pressure forces for the time being. The equation of motion of the outermost particle is then given by

$$\ddot{r} = -GM/r^2, \qquad (4.13)$$

where r is the radius of the blob at any subsequent time.

With the boundary condition that $\dot{r} = 0$, when $r = r_0$, that is that initially the particle starts at rest, equation (4.13) can be integrated to give

$$\dot{r} = -\sqrt{\frac{2GM(r_0 - r)}{rr_0}}. \qquad (4.14)$$

In order to obtain the time of contraction of a blob, this equation has to be integrated from $r = r_0$ to $r = 0$. The equation to be evaluated becomes

$$\int_{r_0}^{0} \sqrt{\frac{r}{r_0 - r}}\, dr = -\sqrt{\frac{2GM}{r_0}}\tau, \qquad (4.15)$$

where τ is the contraction time. Using the substitution $r = r_0 \sin^2 \theta$, the integral can be evaluated and equation (4.15) becomes

$$\sqrt{\frac{2GM}{r_0}}\tau = r_0 \int_0^{\pi/2} (1 - \cos 2\theta)\, d\theta = \frac{\pi r_0}{2}.$$

But $M = 4(\pi \rho_0 r_0^3)/3$, so the contraction time is given by

$$\tau = \sqrt{\frac{3\pi}{32G\rho_0}}. \qquad (4.16)$$

The interesting feature about this equation is that the contraction time depends on no parameter of the cloud except the density. Before we can estimate this contraction time, we need a value for the initial density. If the cloud is to contract as postulated, it must satisfy the condition that the gravitational energy is greater than the thermal energy. This condition has already been expressed as equation (4.3). Written in the way that is of most use in this particular context, it is

$$\rho \geqslant \left(\frac{5\mathscr{R}T}{2G}\right)^3 \frac{3}{4\pi M^2}. \tag{4.17}$$

Even with an estimate for the temperature of such a cloud as low as 10 K, the density cannot be less than about 10^{-15} kg m^{-3} when the cloud separates from the interstellar gas. Inserting this value for ρ_0 into equation (4.16) gives a contraction time of the order of 10^5 years—a very short interval.

Another conclusion that can be reached from equation (4.16) is that the higher the density, the shorter the contraction time. Hence, if a cloud is centrally condensed, as most clouds must surely be in reality, then the denser central regions will contract more quickly. This is the well-known central runaway phenomenon, whose existence has been verified by many detailed computer calculations on cloud collapse. It implies that the centre of a contracting cloud can become dense enough for it to be optically thick, so that radiation can no longer escape except from its surface. The centre effectively becomes star-like, while the outer regions are still in the early stages of contraction and at a low density.

If there is any rotation in the initial cloud, then conservation of angular momentum decrees that this must be present throughout. If the initial state was a dense interstellar gas cloud sharing in the rotation of the Galaxy, then it will possess about 10^{17} m^2 s^{-1} units of angular momentum per unit mass. If angular momentum is conserved

$$h = 10^{17} = r^2\omega, \tag{4.18}$$

where ω is the angular velocity. The rotation becomes important when

$$\frac{GM}{r^2} \sim r\omega^2,$$

or, on using equation (4.18), when

$$r \sim \frac{h^2}{GM}. \tag{4.19}$$

Inserting the appropriate numerical values, we find that account has to be taken of rotational forces when the radius is about 7×10^{13} m, approximately the dimensions of the planetary system. The effect of rotation is obvious: centrifugal force will prevent the cloud from contracting further in the plane of rotation, resulting in the formation of a disk with a young star at its centre (because of the central runaway in the cloud). This result is, of course, very similar to that obtained by Schmidt (1944), who made use of accretion by the already-formed Sun to form the disk. Ter Haar (1948), whose work has been described earlier, was interested mainly in the development of the nebula, but presumably could also have called on such a mechanism for generating it. The formation of this type of nebula was discussed by Hoyle (1949) and Sekiguchi (1961).

Cameron (1962), Schatzmann (1967) and Safranov (1969) are three recent authors who have outlined a scheme basically similar to the above for generating a disk, or solar nebula, surrounding the young Sun. They all obtain a nebula which is somewhat more massive than the sum of the masses of the present planets, usually about $M_\odot/10$. So far, what we have done is to give reasons why the first stage in Kant's theory, namely the formation of a disk surrounding a young star, should occur. We now require a closer investigation of the formation of planets in this disk than Kant's bland statement that secondary condensations form which contract into planets.

The first development is obvious, and has already been described. A temperature gradient is set up in the nebula so that only the non-volatile compounds can condense in its inner region. In this region, grains condense and grow by direct accumulation only. At no stage can the gas be captured by any of these condensations, and so the end result is the formation of terrestrial planet-like objects. (Though, so far as the author knows, no-one has shown that the end result would be four planets with the existing masses, rather than, say, eight planets each of half the present mass, or even forty at a tenth of the present mass). Further away, where the temperature and, hence, the mean velocity of the molecules are lower, and where the disruptive effect of the Sun is less, the growing condensations can also capture the gas, and so form major planets. Again, no exact prediction of the mass and orbital radius of these planets has been made. The discovery of the high-luminosity phase in the evolution of the Sun, which must enhance the temperature gradient, was for a time taken to be indirect corroborative evidence for the correctness of this hypothesis.

A new idea was introduce into the accumulation process by Lyttleton (1956), though it was promptly forgotten by everybody, including Lyttleton, until recently when it has been revived. This is very simply that pressure forces are responsible for holding up the gas in the disk in a direction perpendicular to the plane of rotation. These forces will be much less efficient when acting on grains, and so one might expect the grains to settle towards the rotational plane. Prompted by the work of McCrea & Williams (1965) on the settling of the grains towards the centre of a spherical globe, Schatzmann (1971) postulated that the same mechanism of 'cold welding' of the grains was operative here, and that the grains grew as they settled towards the plane, where they formed a thin disk. Schatzmann's calculations are similar to those given in equations (4.6) to (4.10), but with the gravitational field of a spherical globe replaced by that appropriate for a disk. Schatzmann concluded that a thin carpet of grains can form within the nebula in a short period of time. Lyttleton (1972) has also resurrected his own idea of 1956. He considered the nebula to be somewhat smaller, and he did not think that growth of the grains was necessary during the fall. He argued that grains could reach a size which allowed them to move independently of the gas fairly quickly, and once this occurred the formation of a thin disk of dust was inevitable. Growth into planets is much easier in this dust carpet, as the collision frequency is much higher. But again, it has not yet been shown that the correct number of planets with the correct mass will be formed. Further from the Sun, such condensations can act as nuclei for the capture of hydrogen and helium, leading to the formation of a major planet.

Under the heating influence of the young Sun, much of the gas that had originally been captured will start to evaporate, thus accounting for the difference between the mass of the original nebula and the total mass of the planets. This evaporation will have some effect on the composition of the outer planets. Since all distances are greater there, it takes longer for planets to form, and so, by the time gas is being captured by a planetary condensation, most of the lighter elements hydrogen and helium will have evaporated. This accounts for the outer planets being deficient in these elements.

Detailed calculations regarding the growth of planets have also been carried out by Cameron (1973), Cameron & Pines (1973) and Safranov (1972). Though these authors' theories differ on many points of detail, their motivation is the same. They attempt to follow a grain on to which new material is condensing as it moves through the Solar nebula, and include the possibility that particular compounds can condense on to grains only in different regions of the nebula. Their calculations bear out the general picture outlined above, whereby the grains conform more and more closely to a thin disk as the evolution proceeds, with the rate of growth correspondingly increasing. Safranov (1972) has also attempted to tackle the question of what happens when the grains have grown to a size where they no longer adhere to one another in collision, and have to rely on gravitational capture for further growth. Though these works add considerably to our knowledge of the details of the process of accumulation, they do not significantly change the general picture outlined above.

It is also possible that the asteroids are remnants of this accumulation process, which has now reached some quasi-equilibrium stage whereby destructive collisions play as important a part as collisions where growth by adhering is the outcome. The satellite systems can be explained in terms of secondary accumulations that were captured in orbit about a primary accumulation and did not fall into it. In this way, there would be a tendency for them to be coplanar and prograde. A secondary accumulation that was formed roughly half-way between two planetary accumulations could be captured by either. It could also be captured by one from the other, in which case it would be a prograde satellite for one of them, but a retrograde satellite if captured by the other. One should, therefore, not be surprised at finding retrograde satellites.

It is clear that a feasible theory for the origin of the planets has been outlined, having its basic origins in the work of Kant. The final form of the theory cannot be credited to any individual author, or to any school, since there are a number of variations and because different authors have worked on different aspects of the problem. Since most of the work on this type of theory has been carried out fairly recently, and most of the authors are still active, there is a danger that the argument regarding possible origins of the planets will degenerate into a discussion of which variant of this particular type of theory is most acceptable. Attention should instead concentrate on the wider question of whether this type of theory is better than the best theory in any of the other categories that have been described.

4.4. The Theory of Laplace and its Successors

Laplace (1796) was another of the earlier natural philosophers who proposed a theory for the origin of the planets. According to him, the Sun formed out of a large gas cloud possessing only a small amount of angular momentum. As this cloud contracted, its rotation increased and it took up a lens shape, with the gravitational force balanced by the centrifugal force. Further contraction could now proceed only after the ejection of a ring of material. In time, the further contraction of the Sun rendered it unstable again, with the consequent ejection of another ring. This process of ring ejection continued until the contraction of the Sun ceased at its present radius. Laplace postulated that a planet formed from each of the ejected rings.

The main weakness of Laplace's nebular theory is that we would expect the Sun to be still rotating on the verge of rotational instability. This objection also applies to a modern variant of this theory proposed by Egyed (1960). He placed the problem against the general background of Dirac's cosmology. In this type of cosmology, the universal constants of nature are really variables whose values are determined by the state of the universe. If the universe evolves, then one would expect the value of these universal constants also to be time-dependent. For the actual model under consideration, the gravitational constant, G, is assumed to be inversely proportional to the age of the universe so that its value in the past was greater than it is now. If, therefore, in the distant past the Sun was rotating such that its gravitational attraction was greater than the centrifugal force at the equator, that is

$$\frac{GM}{R^2} > R\omega^2,$$

then the star would be rotationally stable. However, if G is a decreasing function of time, there comes an epoch when the inequality goes the other way, and the Sun becomes rotationally unstable. This leads to the ejection of some matter (taken to be of planetary mass by Egyed) at the equator, which decreases the radius of the Sun, and restores stability. However, G is still decreasing, and so the process repeats itself. The main problem here again appears to be that the Sun should always be rotating on the verge of instability, just waiting for a further decrease in the value of G so that an additional planet can be ejected. A second general objection is that this is a theory of the formation of planetary rings, and not of the formation of planets. The attractive feature about the theory is that it obviously tries to place the origin of the planets in a wider context than just the immediate solar locality. Although it may be argued that Laplace's theory is yet another version of Kant's theory, I believe that the formulation of rings turns it into a different theory which has to be discussed separately.

Berlage (1932; 1934; 1935; 1940; 1948; 1953) has been one of the most prolific of recent writers on the origin of the planets. We have already mentioned his theory and its reliance on electromagnetic effects in another chapter. In the above series of papers, he is concerned with the formation of planets from rings, but his ideas touch on so many other theories that classification becomes difficult. The Sun is assumed to have gathered a nebula about it, the angular momentum being produced by an encounter with another star, as in Schmidt's (1944) theory. The rings are supposed to form in the nebula because all the volume elements follow a Keplerian orbit, and so viscous interactions are set up which dissipate energy. Little is said about the formation of the planets from these rings.

Hoyle (1955) proposed a Laplacian-type theory in qualitative form which was later (Hoyle, 1960) made more quantitative. He assumed that the Sun forms from a rotating gas whose angular momentum is taken to be 4×10^{45} kg m^2 s^{-1} (corresponding to an angular momentum per unit mass of 2×10^{15} m^2 s^{-1}). This is somewhat smaller than the value we have used in the work prior to equation (4.18), because Hoyle maintains that the contracting cloud was magnetically coupled to its surroundings during the initial part of its contraction. With this value, equation (4.18) shows that rotational instability sets in when the radius of the Sun is 3×10^{10} m, almost exactly the present orbital distance of Mercury. At this stage, a ring of material is ejected, allowing the Sun to contract further. Hoyle now deviates from the Laplacian treatment by claiming that the magnetic field of the Sun plays an important part. If the ring is conducting, then it will be coupled to the magnetic field of the Sun, so that angular momentum is transferred to the ring from the Sun. Hoyle finds that, in order to be able to conduct satisfactorily, about 1 atom in 10^7 needs to be ionized, while the solar magnetic field needs to be about 1 gauss.

It should be mentioned that the magnetic coupling method proposed by Hoyle is a variation of an idea proposed by Lüst & Schluter (1955) in a different context. Edgeworth (1962) also pointed out that if the ring was highly conducting, the magnetic field lines would be 'frozen in', thus achieving coupling. Edgeworth equated the proposed mechanism to the torque generated when a metal object is moved through a magnetic field.

If such a coupling exists, then its effect is obvious. As the angular momentum is transferred to the ring, the latter is forced to move away from the Sun to accommodate its higher angular momentum. At the same time, the Sun will be constrained to contract, conserving angular velocity rather than angular momentum. Since the angular velocity is given by h/r^2, after the ring has been ejected the Sun should rotate about 10 times per year, a very reasonable approximation to its present rotation.

The time taken by the ring to move outwards is, of course, governed by the contraction rate of the Sun, and, making use of this, the mean rate of expansion of the ring turns out to be of the order of 10^{-2} m s^{-1}. Now it is the ionized gas that is coupled to the magnetic field, and it is therefore this gas that is driven away from the Sun. Any grains that have condensed are swept along by the gas only if the drag is large enough. Baines et al. (1964) have calculated expressions for the drag on small spheres as

$$R_s = \pi^{3/2} \rho a^2 W \left[\frac{2s}{3} + \frac{2s^3}{15} \right]$$

where a is the radius of the sphere, W the mean thermal velocity of the gas and $s = 2v/(\pi)^{1/2} W$. The grain will be swept along if this force can overcome the gravitational force on the grains. Hoyle shows that grains with a radius in excess of a metre are left behind at the Earth's orbital distance. He therefore concludes that the effect is to form a band of grains in the region where the terrestrial planets are now found. The formation of terrestrial planets thus naturally takes place in this region. Further away from the Sun, similar condensations will form, this time surrounded by gas. It is argued that these condensations capture the gas as well and so form major planets. Near the outer edges of the system, evaporation of hydrogen occurs, and so the outer planets are again deficient in hydrogen.

Hoyle & Wickramasinghe (1968) have updated this theory by taking account of the high-luminosity phase of solar evolution. They conclude that this new feature actually helps to explain the chemical composition of the planets. Fowler, Greenstein & Hoyle (1962) showed that the G, Li, Be, B abundances lead to a requirement that 10 per cent of the terrestrial material must be irradiated with a thermal-neutron flux of 10^{11} nm^{-2} s^{-1} for an interval of 10^7 years. Hoyle's model is capable of meeting these requirements. This theory is capable of explaining most of the features of the solar system, and will be considered with all the other satisfactory theories in the next chapter.

I have not, of course, mentioned all the theories that have been proposed to explain the formation of the planets, but I believe all the main theories have been covered.

5. *Summary and Conclusions*

5.1. Introduction

It is, of course, a very dangerous thing, as history has often shown, to attempt to select the most satisfactory theories for planetary formation, since new observations and new calculations often come along to totally disprove one's arguments. For example, Jeffreys thought in 1918 that he had proved that a tidal type of theory was the only possibility. But half a century later, such theories are very unpopular, whilst monistic theories, where the planets form as part of the process of star formation, are the most generally accepted type. Nevertheless, I believe that some selection can be meaningfully made, though perhaps we cannot narrow the field down to one best theory.

What constitutes an acceptable theory is to a considerable extent a matter of personal taste. Of course, any such theory must be able to explain a large number of the features of the solar system starting from reasonable postulates. But the trouble is that what seems reasonable to some people appears to be totally untenable to others. Worse still, what is 'reasonable' to 'reasonable' people changes with time. Not so long ago, the contention that the World could be any other shape but flat was considered to be unreasonable by all reasonable people, since if it were otherwise the people on the bottom of the Earth would fall off. Who can say what new discovery will change some statement that I have dismissed as being totally unreasonable, and give it the status of being obviously true? Since one can speculate endlessly on which laws of nature will be found to be faulty, or not to apply in particular situations, we must base any judgment we make on the assumption that the laws of nature as we know them today are correct.

Another endless argument is whether all known phenomena must be included in any discussion, if it is to be valid. This argument is often put forward by supporters of theories that make use of magnetic fields. They claim that magnetic fields are known to exist, are known to be very efficient at transporting this or that, and therefore must be included in all discussions of planetary formation. We may compare this with a similar situation in everyday life. Concorde is known to exist (at the time of writing anyway), and is claimed to be a very efficient means of transporting people from one place to another; yet it is totally irrelevant so far as London's internal transport system is concerned. Of course care must be taken not to reverse the argument, and say that because Concorde is irrelevant in transporting Londoners, then it should not be considered as a means of transporting people from London to New York.

In spite of these problems, there are a number of criteria which the theories must satisfy if they are to be acceptable. Without attempting to list all the criteria which readers might think of, we may note:

(1) The number of arbitrary postulates and assumptions must be kept to a minimum.
(2) The general features of the planetary system, listed in Chapter 1, must be accounted for. These features are:
 (a) The Sun is much more massive than the planets.
 (b) The Sun rotates slowly compared with the planets.
 (c) There are about 10 planets in orbit about the Sun.
 (d) The orbits of all these planets are close to a well-defined plane.
 (e) All planets move in a prograde sense along their orbits.
 (f) There are three chemically distinct groups of planets, the terrestrial, the major and the outer.
 (g) The orbital distances of the planets are roughly given by the Titius–Bode Law.
 (h) There are satellites, asteroids, meteorites and comets completing the system.

(3) All the features of the theory must not have a more acceptable explanation incorporated in another theory.
(4) The theory must not predict the existence of features which are not found in the planetary system.

Using these criteria, it is fairly easy to select the theories from the previous three chapters which may be acceptable.

5.2. The Main Theories for the Formation of the Planets

In discussing these theories, we shall preserve the same order as that in which they were described in detail, only now concentrating on their strong and weak points.

Most of the tidal theories clearly fail to satisfy the above criteria on a number of points, the only exception to this blanket condemnation being Woolfson's (1964) theory. If McCrea & Williams' (1965) theory for the formation of the terrestrial planets is incorporated to explain the chemical composition of the planets, then Woolfson's theory can account for all the main features in quite a satisfactory way. Indeed, since actual orbits and masses are calculated for the planets, perhaps it explains more than most theories. It could also be argued that another of its strengths is that it places the formation process at the stage in the evolution of a cluster when stellar encounters are frequent and binary pairs are being formed. Binary stars are very numerous; so this makes Woolfson's background eminently reasonable. However, in precisely this stage lies one of its weaknesses, for it says nothing about star formation, and, if any of the theories that we now have for star formation are correct, then planetary formation appears to be a very likely side result. In that case, stellar encounters become unnecessary. The second weakness of the theory concerns its model of the very young star from which the planetary material is torn. It is assumed that a stage in the stellar contraction exists where the density and temperatures can both have the values selected for them by Woolfson (in order to overcome Spitzer's objection).

Now these are not independent parameters, and since the mass and radius of the star are also selected, both of them should have calculable values, and these may well not be the values adopted by Woolfson. The third weakness in Woolfson's theory is that it is a computer simulation. Now computers work on numerical data only, which means that numerical values have to be selected for every free parameter. The computer then calculates the eventual fate of the system, which turns out to be an excellent reproduction of the planetary system. There is unfortunately no way of knowing what the outcome would be if even one of these numerical values were to be varied, short of re-running the program. One does not know, therefore, whether Woolfson's theory gives a method which is generally capable of forming planets, or whether it works only for a very narrow range of values incorporating those fortuitously chosen by Woolfson.

The theory of Alfvén (1954), as extended by Alfvén & Arrhenius (1970), contains some points common to other theories; for example, it is an accretion theory, so that its initial capture mechanism is similar to other theories of this type. There are three points in the theory which distinguish it from other theories. First of all, there is the use of electromagnetic effects in plasmas to explain the angular momentum distribution in the solar system. Secondly, the planets are thought to have evolved from two distinct nebulae of different compositions, one corresponding to the terrestrial planets, and one to the remaining planets. The need for this double-nebula feature is, of course, tied in with the correctness of the supposition regarding electromagnetic effects in plasmas. The third innovation is the introduction of 'jet streams': the belief that objects on Keplerian orbits tend to become closer and closer to one mean orbit as a consequence of collisions. In order to be mentioned in this section, the theory must be capable of explaining the general features of the planetary system satisfactorily; any strong points to be mentioned must do something over and above this. The obvious such point is that in the jet-stream concept there is a mechanism which can explain many of the features of the asteroids and meteor streams. One of its weaknesses is the same as one in Woolfson's theory; namely, that it does not concern itself with star formation.

and so can take no account of possible automatic planetary formation which follows this event. Though there is no doubt that a simple dipole magnetic field can transfer angular momentum to any infalling plasma in the way described by Alfvén, there is considerable evidence that the present solar wind can carry the magnetic field lines with it, so that, even at large distances, the magnetic field can be approximately *radial*. In the past, with a stronger solar wind, this must have been even more the case. Any accretion-type theory must take account of the solar wind, not only as a deformer of magnetic fields, but also as the generator of a considerable outward force. The inter-collision of a system of particles moving in Keplerian orbits is a very complicated phenomenon, and Alfvén and Arrhenius had to make simplifying assumptions in order to reach their conclusion regarding jet-streaming. There still remains considerable room for doubt as to whether the phenomenon really occurs in practice.

At first sight, the theory of Pendred & Williams (1968) appears to account for most of the solar system, including a tolerable explanation of the masses and orbital distances of the planets. However, the authors were forced to make a great number of simplifying assumptions, and to arbitrarily select several of the numerical values. The good agreement with observation may, therefore, be more a reflection of the skill of the authors in choosing appropriate numerical values than of the intrinsic correctness of the theory. The physics involved is too complicated for it to be possible, at present, to demonstrate in any conclusive way that the theory is correct, and, for this reason, we shall drop it from our list of possibilities. We also drop ter Haar's (1948) theory, not because it is demonstrably incorrect, but rather because every important aspect of the theory somehow seems to be explained more easily by other theories.

From the present point of view, perhaps one of the best things about McCrea's (1960) theory is that it is basically different in most aspects from most other theories. It predicts a solar nebular fragmented into floccules as opposed to the formation of a continuous disk. It predicts that chemical segregation should occur after accumulation into protoplanets, not before. It predicts that protoplanets are formed at low density with a subsequent substantial contraction; whereas, in most theories, planets are accumulated from material not greatly different in density from the present values. Finally, it predicts that the alignment of the planets in a plane is the result of a statistical process, so that there is a real chance that other systems formed in the same manner would not be so well-aligned. Our knowledge concerning most of these points is not quite up to the level where we could use them to distinguish between theories, but we are very close to it. Further details of the possible observations that can be carried out will be given in a following section. The main weakness of the McCrea theory is undoubtedly the concept of a floccule; for this appears to be an object which has too much thermal energy to exist as a separate identity. If it could be shown that such objects do exist in interstellar space, then McCrea's theory should automatically become a front-runner: until this can be done, it must be an also-ran.

The most popular theory at present is some modern form of Kant's theory, different versions of which have been proposed by various authors. It has the major attraction of appearing to make planetary formation an inevitable consequence of star formation. The main weakness is that it is really a theory for the formation of a disk of dust within a nebula. Though it seems reasonable to assume that such a configuration would grow into a small number of massive objects by a combination of collisional accumulation and gravitational accumulation, the theory does not explain why planets should all be so similar in mass (when compensation for composition is made), and it provides no real explanation for planetary spacing (i.e. Bode's law). Some authors have argued, however, that the spacing may not be related to the formation of the planets and that, irrespective of the original distances, the system may eventually evolve to a system with the existing spacings, due to tidal interactions. Dole (1970) and Hills (1970, 1973) have carried out computer simulations which suggest this.

The final theory we mention is that of Hoyle (1960). It suffers from the same weakness as the previous theory, namely that it is a theory for producing a collection of planetisimals, and not for the production of planets. The comment that was made on the theory by Alfvén (1942), namely that the solar wind modifies the magnetic field in the system drastically, is also relevant.

These are the theories that appear to the author to be worthy of further consideration.

5.3. Other Relevant Information

All the above theories satisfy the criterion that they can explain the main features of the planetary system. A number of authors, notably Urey (1956), Latimer (1950) Chamberlain (1952) and Brown (1952) have shown that some of the elements and compounds found on the Earth and in meteorites are not compatible with the Earth having formed from a gas cloud at solar temperatures, while Ringwood (1960) has shown that the thermal properties of the Earth are consistent with it having a cold origin. All the theories mentioned above are consistent with this view and so this cannot be the basis of any selection.

Geological evidence (see Mathers, 1939; Urey, 1966) suggests that the terrestrial planets could well have formed by the aggregation of a large number of small objects. Again all theories have this feature, though in some, (Woolfson, McCrea) these small objects are embedded in a spherical globe, while in the remainder they are within a disk-shaped solar nebula. The density of the surrounding hydrogen is similar in both cases. The work of Anders (1971) placed restrictions on temperature and density in the gas when the planets aggregated, but yet again all our theories are consistent with this view.

It seems that we are not yet in a position to be able definitely to select the one theory which is correct in its description of the process of planetary formation. Is there any reason to believe that the situation will change in the future, or is there any obvious piece of work which can be carried out that will enable us to differentiate between the different theories? The answer is that there are a number of observations which could settle the issue.

Firstly, the discovery of a planetary system containing a significant number of planets which were not particularly well-aligned within the rotational plane of parent star would, I believe, point to the incorrectness of all the theories where the planets form from a flattened rotating nebula. The results of Black & Suffolk (1973) mentioned earlier are therefore of great interest. Confirmation of the observations and analysis are eagerly awaited.

Secondly, the discovery of a planetary system in which the parent star does not, and apparently never did, possess a magnetic field would invalidate the theories of Alfvén and Hoyle.

Observations of the interstellar medium continually give us more information regarding interstellar clouds. It is now recognized that many clouds are fragmented into a vast number of subclouds. If any of these subclouds were found to be gravitationally unstable in the traditional sense, this would lend support to the floccule hypothesis of McCrea.

Finally, Walker (1956, 1959) and others have studied young stellar clusters in some detail. Walker (1972) believes that some young stars are accreting material in a discrete form, which might add support to McCrea's theory.

We cannot, therefore, choose the correct theory for the origin of the planets at present, though important clues may become available in the near future. We can only speculate and make our choice on purely personal grounds between the five theories that have survived all the tests.

5.4. Concluding Remarks

On a personal level, it seems to the author that if a theory for the formation of stars also accounts for planetary formation, then any other mechanism for the formation of planets, though theoretically possible, becomes unnecessary. Measured by this yardstick, there remain only two types of theories for immediate consideration, though, if they are both subsequently proved to be incorrect, then other theories will have to be reconsidered. The two theories are the floccule theory of McCrea and the group of theories following Kant, associated with the names of Cameron, Safranov, and Schatzmann.

As has already been mentioned, McCrea's theory is very weak in its initial stages, but is subsequently well worked out, whereas the Kant-type theory seems very plausible until the final stages of accumulation into planets. It is a pity that no way has yet been found to combine these in some way so that the strongest points of both types of theories survive.

In conclusion, though I have attempted to be objective throughout, it should be realized that it is almost inevitable that personal bias has entered into the discussion somewhere and that justice has not been done to some theory. Perhaps the author of any such theory can take comfort from the fact that, if history provides any indication, a theory now labelled by the majority of astronomers as 'unlikely', 'unworkable' or any similar adjective, will no doubt find itself the most popular choice in a few decades.

References

Alfvén, H. (1942*a*). *Stockh. Obs. Ann.*, **14**, No. 2.
Alfvén, H. (1942*b*). *Stockh. Obs. Ann.*, **14**, No. 5.
Alfvén, H. (1943). *Nature*, **152**, 221.
Alfvén, H. (1946). *Stockh. Obs. Ann.*, **14**, No. 9.
Alfvén, H. (1954). *On the origin of the Solar System*, Oxford University Press.
Alfvén, H. (1969). *Astrophys. Sp. Sci.*, **4**, 84.
Alfvén, H. (1970). *Astrophys. Sp. Sci.*, **6**, 161.
Alfvén, H. (1971). *Science*, **173**, 522.
Alfvén, H. & Arrhenius, G. (1970*a*). *Astrophys. Sp. Sci.*, **8**, 338.
Alfvén, H. & Arrhenius, G. (1970*b*). *Astrophys. Sp. Sci.*, **9**, 3.
Alfvén, H. & Arrhenius, G. (1973). *Astrophys. Sp. Sci.*, **21**, 117.
Allen, C. W. (1973). *Astrophysical Quantities*, 3rd ed., Athlone Press.
Anders, E. (1971). *Ann. Rev. Astron. Astrophys.*, **9**, 1.
Anders, E. (1972). *Nice Symposium on the Origin of the Solar System*, C.N.R.S., 179.
Arrhenius, S. (1913). *Das Werden der Welten*, Leipzig.
Aust, C. & Woolfson, M. M. (1971). *Mon. Not. R. Astr. Soc.*, **153**, 21P.
Baines, M. J., Williams, I. P. & Asebiomo, A. S. (1965). *Mon. Not. R. Ast. Soc.*, **130**, 63.
Banerji, A. C. & Strivastava, K. M. (1963). *Proc. Natn. Acad. Sci. India*, **A33**, 125.
Berlage, H. P. (1927). *Gerlands Beitr. Geophys*, **17**.
Berlage, H. P. (1930*a*). *Proc. K. Ned. Akad. Wet.*, **33**, 614 and 719.
Berlage, H. P. (1930*b*). *Het. Ontstaan en Vergaan der Werelden*, Amsterdam.
Berlage, H. P. (1932). *Proc. K. Ned. Akad. Wet.*, **35**, 553.
Berlage, H. P. (1934). *Proc. K. Ned. Akad. Wet.*, **37**, 221.
Berlage, H. P. (1935). *Proc. K. Ned. Akad. Wet.*, **38**, 857.
Berlage, H. P. (1940). *Proc. K. Ned. Akad. Wet.*, **43**, 532.
Berlage, H. P. (1948). *Proc. K. Ned. Akad. Wet.*, **51**, 796 and 965.
Berlage, H. P. (1953). *Proc. K. Ned. Akad. Wet.*, **56**, 456.
Berlage, H. P. (1959). *Proc. K. Ned. Akad. Wet.*, **B62**, 63 and 73.
Berlage, H. P. (1962). *Proc. K. Ned. Akad. Wet.*, **B65**, 211.
Bickerton, A. W. (1878). *Trans. N.Z. Inst.*, **11**, 125.
Birkeland, K. (1912). *C.R. Hebd. Séanc. Acad. Sci.*, **155**, 892.
Black, D. C. & Suffolk, G. C. J. (1973). *Icarus*, **19**, 353.

Bondi, H. (1952). *Mon. Not. R. Ast. Soc.*, **112**, 195.
Brown, H. (1952). *The Atmospheres of the Earth and Planets*, ed. Kuiper, Chicago University Press.
Buffon, G. L. L. (1745). *De la Formation des Planetes*, Paris.
Cameron, A. G. W. (1962). *Icarus*, **1**, 18.
Cameron, A. G. W. (1973). *Icarus*, **18**, 407.
Cameron, A. G. W. & Pines, M. R. (1973). *Icarus*, **18**, 377.
Chamberlain, T. C. (1901). *Ap. J.*, **14**, 17.
Chamberlain, T. C. (1927). *The Origin of the Earth*, Chicago University Press.
Chamberlain, T. C. (1952). *The Atmosphere of the Earth and Planets*, Chicago University Press.
Cox, J. P. & Guili, R. T. (1968). *Principles of Stellar Structure*, Gordon and Breach.
Cremin, A. W. (1969). *Ph.D. Thesis*, University of Reading.
Dauvillier, M. A. (1942a). *Archs. Sci. Phys. Nat.*, **24**, 5.
Dauvillier, M. A. (1942b). *Archs. Sci. Phys. Nat.*, **24**, 65.
Dermott, S. F. (1972). *Nice Symposium on the Origin of the Solar System*, C.N.R.S., 320.
Descartes, R. (1644). *Principia Philosphiae*, Amsterdam.
Dole, S. H. (1970). *Icarus*, **13**, 494.
Donnison, J. R. & Williams, I. P. (1974). *Astrophys. Sp. Sci.*, **29**, 387.
Dormond, J. R. & Woolfson, M. M. (1971). *Mon. Not. R. Ast. Soc.*, **151**, 307.
Edgeworth, K. E. (1946). *Mon. Not. R. Ast. Soc.*, **106**, 473.
Edgeworth, K. E. (1949). *Mon. Not. R. Ast. Soc.*, **109**, 600.
Edgeworth, K. E. (1962). *Observatory*, **82**, 219.
Egyed, L. (1960). *Nature*, **186**, 621.
Ezer, D. & Cameron, A. G. W. (1963). *Icarus*, **1**, 422.
Faulkner, J., Griffiths, K. & Hoyle, F. (1963). *Mon. Not. R. Ast. Soc.*, **126**, 1.
Fowler, W. A., Greenstein, J. & Hoyle, F. (1962). *Geoph. J. R. Ast. Soc.*, **6**, 148.
Gunn, R. (1932). *Phys. Rev.*, **38**, 311.
Hayashi, C. (1961). *Pub. Ast. Soc. Japan*, **13**, 450.
Hellyer, B. (1970). *Mon. Not. R. Ast. Soc.*, **143**, 383.
Henyey, L. G., Le Levier, R. & Levee, R. D. (1955). *Pub. A.S.P.*, **67**, 4.
Hills, J. G. (1970). *Nature*, **225**, 840.
Hills, J. G. (1973). *Icarus*, **18**, 505.
Hoyle, F. (1944). *Proc. Camb. Phil. Soc.*, **40**, 256.
Hoyle, F. (1945). *Mon. Not. R. Ast. Soc.*, **105**, 175.
Hoyle, F. (1946a). *Mon. Not. R. Ast. Soc.*, **106**, 343.
Hoyle, F. (1946b). *Mon. Not. R. Ast. Soc.*, **106**, 406.
Hoyle, F. (1955). *Frontiers of Astronomy*, Heinemann.
Hoyle, F. (1960). *Qrt. J. R. Ast. Soc.*, **1**, 28.
Hoyle, F. & Lyttleton, R. A. (1939). *Proc. Camb. Phil. Soc.*, **36**, 424.
Hoyle, F. & Wickramasinghe, W. C. (1968). *Nature*, **217**, 415.

Jeans, J. (1916). *Mon. Not. R. Ast. Soc.*, **77**, 84.
Jeans, J. (1917a). *Mon. Not. R. Ast. Soc.*, **77**, 186.
Jeans, J. (1917b). *Mem. R. Ast. Soc.*, **62**, 1.
Jeans, J. (1928). *Astronomy and Cosmogony*, Cambridge University Press.
Jeans, J. (1931). *Nature*, **128**, 432.
Jeffreys, H. (1918). *Mon. Not. R. Ast. Soc.*, **78**, 424.
Jeffreys, H. (1924). *The Earth*, Cambridge University Press.
Jeffreys, H. (1929). *Mon. Not. R. Ast. Soc.*, **89**, 636.
Kant, I. (1755). *Algemeine Naturgeschichte und Theories des Himmels*.
Kerridge, J. F. & Vedder, J. F. (1972). *Nice Symposium on the Origin of the Solar System*, C.N.R.S., 282.
Kramers, J. & Burgers, J. M. (1946). *Proc. K. Ned. Akad. Wet.*, **49**, 589.
Kuhi, L. V. (1964). *Ap. J.*, **140**, 1409.
Kuiper, G. (1949). *Ap. J.*, **109**, 308.
Kuiper, G. (1951a). *Proc. Nat. Acad. Sci. U.S.A.*, **3**, 1.
Kuiper, G. (1951b). *Astrophysics*, ed. Hynek, McGraw-Hill.
Laplace, P. S. (1796). *Exposition du systeme de Monde*, Paris.
Latimer, W. M. (1950). *Science*, **112**, 101.
Levin, B. J. (1958). *Origin of the Earth and Planets*, Moscow.
Lüst, R. & Schluter, A. (1955). *Z. Astrophys.*, **38**, 190.
Luyten, W. J. (1933). *Z. Astrophys.*, **1**, 46.
Lyttleton, R. A. (1936). *Mon. Not. R. Ast. Soc.*, **96**, 559.
Lyttleton, R. A. (1937). *Mon. Not. R. Astr. Soc.*, **97**, 633.
Lyttleton, R. A. (1938a). *Mon. Not. R. Ast. Soc.*, **98**, 536.
Lyttleton, R. A. (1938b). *Mon. Not. R. Astr. Soc.*, **98**, 646.
Lyttleton, R. A. (1940). *Mon. Not. R. Astr. Soc.*, **100**, 546.
Lyttleton, R. A. (1941a). *Mon. Not. R. Astr. Soc.*, **101**, 216.
Lyttleton, R. A. (1941b). *Mon. Not. R. Astr. Soc.*, **101**, 349.
Lyttleton, R. A. (1956). *The Mystery of the Comets*, Cambridge University Press.
Lyttleton, R. A. (1960). *Mon. Not. R. Astr. Soc.*, **121**, 551.
Lyttleton, R. A. (1961). *Mon. Not. R. Astr. Soc.*, **122**, 399.
Lyttleton, R. A. (1972). *Mon. Not. R. Astr. Soc.*, **150**, 463.
Mathers, K. F. (1939). *Science*, **89**, 65.
McCrea, W. H. (1957). *Mon. Not. R. Astr. Soc.*, **117**, 562.
McCrea, W. H. (1960a). *Ciel et Terre*, **76**, 11.
McCrea, W. H. (1960b). *Proc. Roy. Soc.*, **A256**, 245.
McCrea, W. H. (1963). *Contemp. Phys.*, **4**, 278.
McCrea, W. H. & Williams, I. P., (1965). *Proc. Roy. Soc.*, **A287**, 143.
Moulton, F. R. (1905). *Ap. J.*, **22**, 165.
Nieto, M., (1973). *Titius–Bode Law of Planetary Distance*, Pergamon.
Nolke, F. (1930). *Der Entwicklungsgang Unseres Planeten Systems*, Bonn.
Nolke, F. (1932). *Mon. Not. R. Astr. Soc.*, **93**, 159.
Pendred, B. W. & Williams, I. P. (1968). *Icarus*, **8**, 129.

Pendred, B. W. & Williams, I. P. (1969). *Astrophys. Sp. Sci.*, **5**, 420.
Prentice, A. J. R. & ter Haar, D. (1969). *Acta. Phys. Hung.*, **27**, 231.
Ringwood, A. E. (1960). *Geochim. Cosmochim Acta*, **20**, 241.
Richter, N. B. (1963). *The Nature of Comets*, Methuen.
Russell, H. N. (1935). *The Solar System and its Origin*, Macmillan, New York.
Safranov, V. S. (1969). *Evolution of the Preplanetary Cloud and the formation of the Earth and Planets*, Moscow.
Safranov, V. S. (1972). *Nice Symposium on the Origin of the Solar System*, C.N.R.S., 89.
Sarvajna, D. K. (1970). *Astrophys. Sp. Sci.*, **6**, 258.
Schatzmann, E. (1967). *Ann. Astroph*, **30**, 963.
Schatzmann, E. (1971). *Physics of the Solar System*, Goddard Inst. X630-71-3.
Schmidt, O. Y. (1944). *Dokl. Akad. Nauk S.S.S.R.*, **45**, No. 6.
Schwarzchild, M. (1958). *Structure and Evolution of the Stars*, Princeton University Press.
See, T. J. J. (1910). *The Capture Theory of Cosmical Evolution*, Nicholas Lynn, Mass.
Sekiguchi, N. (1961). *Pub. Astron. Soc. Japan*, **13**, 241.
Spitzer, L. (1939). *Astrophys. J.*, **90**, 675.
ter Haar, D. (1948). *Kgl Danske Videnscap Selskab. Nat-Fys. Medd*, **25**, 3.
ter Haar, D. (1950). *Astrophys. J.*, **111**, 179.
Urey, H. C. (1956). *Astrophys. J.*, **124**, 623.
Urey, H. C. (1966). *Mon. Not. R. Astr. Soc.*, **135**, 199.
van de Kamp, P. (1969*a*). *Pub. A.S.P.*, **81**, 5.
van de Kamp, P. (1969*b*). *Astron. J.*, **74**, 757.
van de Kamp, P. (1971). *Ann. Rev. Astron. and Astrophys.*, **9**, 103.
von Weizsäcker, C. F. (1944). *Z. Astrophys.*, **22**, 319.
Walker, M. F. (1956). *Ap. J. Supp.*, **2**, 365.
Walker, M. F. (1959). *Ap. J.*, **130**, 57.
Walker, M. F. (1972). *Private Communication.*
Whipple, F. L. (1946). *Astrophys. J.*, **104**, 1.
Whipple, F. L. (1948). *Harv. Obs. Monog.*, No. 7.
Williams, I. P. (1967). *Mon. Not. R. Astr. Soc.*, **136**, 341.
Williams, I. P. (1969). *Mon. Not. R. Astr. Soc.*, **146**, 339.
Williams, I. P. (1971). *Astrophys. Sp. Sci.*, **12**, 165.
Williams, I. P. (1972). *Nice Symposium on the Origin of the Solar System*, C.N.R.S. 274.
Williams, I. P. (1972). *Astrophys. Sp. Sci.*, **18**, 223.
Williams, I. P. (1974). *Exploration of the Planetary System*, I.A.U., 3.
Williams, I. P. & Crampin, D. J. (1971). *Mon. Not. R. Astr. Soc.*, **152**, 261.
Williams, I. P. & Cremin, A. W. (1968). *Qrt. J. R. Astr. Soc.*, **9**, 40.
Williams, I. P. & Cremin, A. W. (1969). *Mon. Not. R. Astr. Soc.*, **144**, 359.
Williams, I. P. & Cremin, A. W. (1970). *Les Congres et Colloques de L'Université de Liege*, **59**, 147.

Williams, I. P. & Donnison, J. R. (1973). *Mon. Not. R. Astr. Soc.*, **165**, 295.
Williams, I. P. & Galley, S. J. (1971). *Mon. Not. R. Astr. Soc.*, **151**, 207.
Williams, I. P. & Handbury, M. J. (1974). *Astrophys. Space Sci.*, **30**, 215.
Woolfson, M. M. (1960). *Nature*, **187**, 147.
Woolfson, M. M. (1964). *Proc. Roy. Soc.*, **A282**, 485.
Woolfson, M. M. (1969). *Prog. Phys.*, **32**, 135.

Subject Index

Accretion 51, 92
Angular momentum 61, 73, 85
    conservation 87
    of planets 6
    of Sun 6
Ariel 19
Asteroids 15, 20, 84

Barnard's star 4, 5, 96
Binary stars 3, 42

Callisto 19
α Centauri 3
Ceres 20
Chemical composition 7, 17, 23, 24, 49, 55
Chondrites 23, 24
Chondrules 23
Contraction time 13
Convection 10
Cosmic rays 68
61 Cygni 4, 5

Deimos 19
Dione 19

Earth 14, 15, 23, 44
Eccentricity 15, 19
Ellipsoid 31
Enceladus 19

Energy, gravitational 7, 12, 40, 71
    thermal 40, 71
ε Eridani 4, 5
Escape velocity 51
Europa 19

Filament formation 37
Floccules 71

Ganymede 19
Grains 49, 55, 75, 83, 88

HR diagram 9
Hyperon 19

Iapetus 19
Inclination 15, 19, 20
Instability 36
Io 19

Janus 19
Jet stream 60, 92
Jupiter 3, 6, 14, 15

Kepler's law 3
Kramer's law 8

Lalande 4, 5
Luminosity 6

Magnetic field 7, 56, 68, 87
Major planets 17
Mars 14, 15
Mass, distribution 27
    solar 6
    planets 15
Mercury 12, 14, 15
Meteorites 23
Mimas 19
Minor planets 15, 20
Miranda 19
Moon 19, 44

Nebula 30
    solar 54, 56, 68, 81
Neptune 14, 15
Nereid 19
Novae 29, 45

Oberon 19
Opacity 8
Optical depth 9
Orbital elements 15
Outer planets 17

Pallas 20
Parallax 4
Phobos 19
Phoebe 19
Planetary system 25
    others 2
Planetesimals 30
Planets, data 14, 15
    major 17
    outer 17
    terrestrial 17
Pluto 14, 15
Polytrope 8, 10
Pressure 9
Prograde orbit 18, 74
Protoplanets 69, 73
Proto-Sun 30

Radiation transfer 7
Retrograde orbits 18, 73
Rhea 19
Rotation 6, 16, 36

Satellites 18
Saturn 14, 15
Segregation 75
Silicates 23
Solar, nebula 54, 56, 68, 81
    wind 7, 12, 22, 78
Star, Barnard 4, 5, 96
    Binary 42
    T-Tauri 12
Sun 6
Supernovae 45

T-Tauri stars 12
Temperature 7, 9, 25, 40, 46, 54, 71
Terrestrial planets 17
Tethys 19
Theories, classification 12
Tidal effects 29, 49, 77
Titan 19
Titania 19
Titius–Bode law 14, 20, 44, 67, 90
Triton 19
Turbulence 68, 70

Umbriel 19
Uranus 14, 15

Venus 14, 15
Viscosity 75
Vortex 66

Wind, solar 7, 12, 22, 78

Author Index

Alfvén 57, 60, 92, 95
Allen 7, 15, 16, 19
Anders 24, 95
Arrhenius, G. 60, 92
Arrhenius, S. 30
Asebiomo 76
Aust 74

Baines 76
Banerji 45
Berlage 56, 86
Bickerton 29
Birkeland 56
Black 4, 5, 26, 96
Bondi 51, 53
Brown 95
Buffon 29
Burgers 41

Cameron 11, 82, 84, 96
Chamberlain 30, 56, 95
Cox 8, 12
Crampin 76
Cremin 28, 46, 51, 53, 73

Dauvillier 44
Dermott 18
Descartes 65, 78
Dole 94
Donnison 74, 78
Dormond 49

Edgeworth 69, 87
Egyed 85
Ezer 11

Faulkner 11
Fowler 88

Galley 74
Greenstein 88
Griffiths 11
Guili 8, 12
Gunn 44

Handbury 76, 78
Hayashi 11
Hellyer 21
Henyey 8
Hills 94
Hoylc 11, 44, 45, 51, 52, 82, 87, 88, 94

Jeans 28, 31, 36, 37, 47
Jeffreys 30, 36, 37, 70, 89

Kant 65, 79, 96
Kerridge 75
Kramer 8, 41
Kuhi 12
Kuiper 69

Laplace 65, 85
Latimer 95
Le Levier 8
Levee 8
Levin 101
Lüst 87
Luyten 38
Lyttleton 22, 40, 43, 51, 52, 54, 83

Mathers 95
McCrea 28, 49, 70, 71, 75, 77, 83, 91, 93
Moulton 30, 56

Nieto 14
Nolke 38, 40, 43, 44

Pendred 60, 63, 93
Pines 84
Prentice 68

Richter 22
Ringwood 95
Russell 38, 40, 43, 46

Safranov 82, 84, 96
Sarvajna 40
Schatzmann 82, 83, 96
Schluter 87
Schmidt 54, 56, 66, 82, 86
Schwarzschild 8
See 30
Sekiguchi 82
Spitzer 38, 40, 42, 44, 91
Strivastava 45
Suffolk 4, 5, 26, 96

ter Haar 68, 82, 93

Urey 78, 95

van de Kamp 4, 5
Vedder 75
von Weizsäcker 66

Walker 96
Whipple 69
Wickramasinghe 88
Williams 12, 28, 46, 49, 60, 63, 72, 73, 74, 75, 76, 78, 83, 91, 93
Woolfson 45, 46, 49, 74, 77, 91